Springer Series in Reliability Engineering

Series Editor

Professor Hoang Pham
Department of Industrial and Systems Engineering
Rutgers, The State University of New Jersey
96 Frelinghuysen Road
Piscataway, NJ 08854-8018
USA

Other titles in this series

The Universal Generating Function in Reliability Analysis and Optimization
Gregory Levitin

Warranty Management and Product Manufacture
D.N.P Murthy and Wallace R. Blischke

Maintenance Theory of Reliability
Toshio Nakagawa

System Software Reliability
Hoang Pham

Reliability and Optimal Maintenance
Hongzhou Wang and Hoang Pham

Applied Reliability and Quality
B.S. Dhillon

Shock and Damage Models in Reliability Theory
Toshio Nakagawa

Risk Management
Terje Aven and Jan Erik Vinnem

Satisfying Safety Goals by Probabilistic Risk Assessment
Hiromitsu Kumamoto

Offshore Risk Assessment (2nd Edition)
Jan Erik Vinnem

The Maintenance Management Framework
Adolfo Crespo Márquez

Human Reliability and Error in Transportation Systems
B.S. Dhillon

Complex System Maintenance Handbook
D.N.P. Murthy and K.A.H. Kobbacy

Recent Advances in Reliability and Quality in Design
Hoang Pham

Product Reliability
D.N. Prabhakar Murthy, Marvin Rausand and Trond Østerås

Mining Equipment Reliability, Maintainability, and Safety
B.S. Dhillon

Toshio Nakagawa

Advanced Reliability Models and Maintenance Policies

Springer

Toshio Nakagawa, Dr. Eng.
Dept. of Marketing and Information Systems
Aichi Institute of Technology
1247 Yachigusa
Yakusa-cho
Toyota 470-0392
Japan

ISBN 978-1-84996-772-3   e-ISBN 978-1-84800-294-4

DOI 10.1007/978-1-84800-294-4

Springer Series in Reliability Engineering ISSN 1614-7839

British Library Cataloguing in Publication Data
Nakagawa, Toshio, 1942-
  Advanced reliability models and maintenance policies. -
  (Springer series in reliability engineering)
  1. Reliability (Engineering)
  I. Title
  620'.00452

Cover design: deblik, Berlin, Germany

Printed on acid-free paper

9 8 7 6 5 4 3 2 1

springer.com

Preface

Many accidents have happened this year in the whole world: A big earthquake in Japan caused heavy damage to a nuclear power plant. An automobile company that produces a small but important part stopped its operation due to the earthquake. This shut down all automobile companies for several days because of the lack of parts. A free way bridge in the US fell into a river. Most such serious matters might be prevented if practical methods of reliability and maintenance were correctly and effectively used.

I have already published the first monograph *Maintenance Theory of Reliability* [1] that summarizes maintenance policies for system reliability models and the second monograph *Shock and Damage Models in Reliability Theory* [3] that introduces reliability engineers to kinds of damage models and their maintenance policies. Since then, our research group in Nagoya has continued to study reliability and maintenance theoretically and to apply them practically to some actual models. Some useful methods may be applicable to other fields in management science and computer systems.

Reliability becomes of more concerns to engineers and managers engaged in making high quality products, designing reliable systems, and preventing serious accidents. This book deals primarily with a variety of advanced stochastic and probability models related to reliability: Redundancy, maintenance, and partition techniques can improve system reliability, and using these methods, various policies that optimize some appropriate measures of management and computer systems are discussed analytically and numerically.

This book is composed of ten chapters: From the viewpoints of maintenance, we take up maintenance policies for a finite time span in Chap. 4 and replacement policies with continuous and discrete variables in Chap. 8. Furthermore, Chap. 5 has another look at reliability models from the points of forward time to the future and backward time to the past. Next, referring to the classification of redundancy, we summarize the reliability properties of some parallel systems and redundant models of data transmission in Chap. 2, and discuss optimum policies of retries as recovery techniques of computer systems in Chap. 6 and of checkpoint schemes for fault detection database systems in Chap. 7. Finally, we apply some optimization problems in management science in Chap. 10 by using reliability techniques.

New subjects of study such as traceability, system complexity, service reliability, and entropy model are proposed. A golden ratio for the first time in reliability is presented twice. It is of interest in partition policies that the summation of integers from 1 to N plays an important role in solving optimization problems with discrete variables. In addition, the contrasts with forward and backward times, redundancy and partition, and continuous and discrete variables presented in this book are like a study in opposites. These will be helpful for graduate students and researchers who search for new themes of study in reliability theory, and for reliability engineers who engage in maintenance work. Furthermore, many applications to management science and computer systems are useful for engineers and managers who work in banks, stock companies, and computer industries.

I wish to thank Professor Kazumi Yasui, Dr. Mitsuhiro Imaizumi, and Dr. Mitsutaka Kimura for Chaps. 2, 3 and 6, Professor Masaaki Kijima for Chap. 4, Professor Hiroaki Sandoh for Chaps. 5 and 6, Dr. Kenichiro Naruse and Dr. Sayori Maeji for Chap. 7, Dr. Kodo Ito for Chap. 8, and Professor Shouji Nakamura for Chap. 10, who are co-workers on our research papers. I wish to express my special thanks to Dr. Satoshi Mizutani, Mr. Teruo Yamamoto, Mr. Shigemi Osaki, Dr. Kazunori Iwata and his wife Yorika for their encouragement and support in writing and typing this book. Finally, I would like to give my sincere appreciation to Professor Hoang Pham, Rutgers University, and editor Anthony Doyle, Springer-Verlag, London, for providing the opportunity for me to write this book.

Toyota, Japan   *Toshio Nakagawa*
December 2007

Contents

1

Introduction

The importance of maintenance and reliability will be greatly enhanced by environmental considerations and the protection of natural resources. Maintenance policies and reliability techniques have to be developed and expanded as proposed models become more complex and large-scale. They also will be applied not only to daily life, but also to a variety of other fields because consumers, workers, and managers must make, buy, sell, use, and operate articles and products with a sense of safety and security.

In the past five decades, valuable contributions to reliability theory have been made. The first book [1] was intended to summarize our research results in the past four decades based on the book [2]: Standard policies of repair, replacement, preventive maintenance, and inspection were taken up, and their optimum policies were summarized in detail. The second book [3] introduced damage models by using stochastic processes, discussed their maintenance policies, and applied their results to computer systems.

The first book could not embody some results of advanced reliability models because we restricted ourselves mainly to the basic ones. After that, our research group obtained new results and is searching for advanced stochastic models in other fields of reliability by using reliability techniques, and conversely, for analyzing reliability models by using useful techniques in other fields.

We aim to write this book from the reliability viewpoints of maintenance, redundancy, and applications. Finally, we describe briefly further study presented in this book and suggest establishing *Anshin Science* from the reliability viewpoint in the near future. Anshin is a Japanese word that includes all meanings of safety, security, and assurance in English. I hope that the world of *Anshin* would be spread all over the world.

1.1 Maintenance

The number of aged plants such as chemical, steel and power plants has increased greatly in Japan. For example, about one-third of such plants are now from 17 to 23 years old, and about a quarter of them are older than 23 years. Recently, some houses were on fire from the origin in old electric appliances. Especially, Japan is subject to frequent earthquakes. Big earthquakes happened this year and the last year and inflicted serious damage on an automobile manufacturer and a nuclear power plant. This causes a sense of social instability and exerts an unrecoverable bad influence on the living environment. Furthermore, public infrastructures in advanced nations have grown old and will be unserviceable in the near future. A freeway bridge in the US fell suddenly into a river the last year with tragic results. In addition, big typhoons, hurricanes, and cyclones are born frequently all over the world and incur heavy losses to the nations affected. From such viewpoints of reliability, *maintenance* will be highly important in a wide sense, and its policies should be properly and quickly established.

In the past five decades, valuable contributions to maintenance theory in reliability have been developed. The first book [1] summarized the research results studied mainly by the author in the past four decades based on the book [2]: Standard policies for maintenance of repair, replacement, preventive maintenance, and inspection were collected, and their optimum policies were discussed heatedly. Moreover, the second book [3] introduced maintenance policies for damage models, using stochastic processes, and applied their results to the analysis of computer systems. A survey of theories and methods of reliability and maintenance on multi-unit systems and their current hot topics were presented [4]. The first book could not embody some advanced research results because we restricted ourselves mainly to the basic ones. In the past 10 years, our research group obtained new interesting results.

There have few papers on maintenance for a finite time span because it is more difficult theoretically to analyze their optimization. We obtain optimum periodic policies of inspection and replacement in Chap. 3, using the partition method, and sequential policies of imperfect preventive maintenance(PM), inspection and cumulative damage models in Chap. 4. It is shown that optimum maintenance times are given by solving simultaneous equations numerically. We deal with replacement and PM models with continuous and discrete variables in Chap. 8. The computing procedures for obtaining optimum policies are specified. Most models considered treat fitted maintenance for operating units that will fail in the near future, that is called the *forward time model*. However, when a unit is detected to have failed and its failure time is unknown, we often know when it failed, that is called the *backward time model*. In Chap. 5, we investigate the properties of forward and backward times and apply them to the maintenance of backward time models in a database system and the reweighing of a scale.

1.2 Redundancy

High system reliability can be achieved by redundancy. A classical problem is to determine how reliability can be improved by using redundant units. The results of various redundant systems with repair were summarized as the *repairman problem*. Optimization problems of redundancy and allocation subject to some constraints were solved, and qualitative relationships were obtained for multi-component structures [2]. Furthermore, some useful expressions of reliability measures of many redundant systems were shown [5,6]. The fundamentals and applications of system reliability and reliability optimization in system design were well described [7]. Various combinatorial optimization problems with multiple constraints for different system structures were considered [8], and their computational techniques were surveyed [9].

Transient and intermittent faults occur in a computer system, and sometimes, take the form of errors that lead to system failure. Three different techniques for decreasing the possibility of fault occurrences can be used [10]: *Fault avoidance* is preventing fault occurrences by improving qualities of structural parts and placing them well in their surroundings. *Fault masking* is preventing faults by error correction codes and majority voting. *Fault tolerance* is that a system continues to function correctly in the presence of hardware failures and software errors. These techniques above are called simply fault tolerance.

Redundant techniques of a system for improving reliability and achieving fault tolerance are classified commonly into the following forms [10–13]:

(1) Hardware Redundancy
   (a) *Static hardware redundancy* is an error masking technique in which the effects of faults are essentially hidden from the system with no specific indication of their occurrence. Existing faults are not removed. A typical example is a triple modular redundancy.
   (b) *Dynamic hardware redundancy* is a fault tolerance technique in which the system continues to function by detection and removing faults, replacing faulty units, and making reconfigurations. Typical examples are standby sparing systems and graceful degrading systems [14].
   (c) *Hybrid hardware redundancy* is a combination of the advantages of static and dynamic hardware redundancies.
(2) Software Redundancy
   This technique uses extra codes, small routines or possibly complete programs to check the correctness or consistency of the results produced by software. Typical examples are N-version programming and Ad-Hoc techniques.
(3) Information Redundancy
   This technique adds redundant information to data to allow fault detection, error masking, and fault tolerance. Examples are error-detecting codes such as parity codes, signatures, and watchdog processors.
(4) Time Redundancy
   This technique involves the repetition of a given computation a number

of times and the comparison of results. This is used to detect transient or intermittent faults, to mask errors, and to recover the system. Typical examples are retries and checkpoint schemes.

Redundancies (1), (2), and (3) are also called *Space Redundancy* because high reliability is attained by providing multiple resources of hardware and software.

Referring to the above classification, we take up a variety of optimization problems encountered in the field of redundancy. In Chap. 2, from the viewpoint of hardware redundancy, we summarize our research results for a parallel system that is the most standard redundant system and investigate the properties of series-parallel and parallel-series system. We also compare three redundant systems with the same mean failure time. In addition, we define the complexity of redundant systems as the number of paths and the entropy. As practical examples of information and time redundancies, we propose three models of data transmission and redundant models with bits, networks, and copies. From the viewpoint of time redundancy, in Chap. 6, we solve the optimization problem of how many number of retrials should be done when the trial of some event fails. In Chap. 7, we take up several checkpoint schemes for redundant modular systems as recovery techniques.

On the other hand, reliability models exist whose performance may improve by partitioning their functions suitably. In Chap. 3, we specify the optimum partition method and obtain optimum policies for maintenance models and computer systems.

1.3 Applications

We have had a look at which techniques and maintenance policies grown in reliability theory give full benefit in the fields of computer, information, and communication systems. Using the theory of cumulative processes in stochastic processes [15–17], we have applied cumulative damage models to the garbage collection of a computer system and the backup scheme of a database system [18]. Furthermore, we have already analyzed a storage system such as missiles [19], a phased array radar system [20], a FADEC (Full Authority Digital Engine Control) system [21], and a gas turbine engine of cogeneration system [22], using the techniques of maintenance and reliability.

It has been well-known that the theory of Martingale and Brownian motion in stochastic processes contributes greatly to mathematical analysis of finance [23, 24]. Similarly, the methods and results of reliability theory are useful for solving optimization problems in management science. In Chap. 10, we first define service reliability theoretically and investigate its properties. Next, we consider the number of spare cash boxes for unmanned ATMs in a bank and their maintenance. Furthermore, we determine an adequate loan interest rate, introducing the probabilities of bankruptcy and mortgage collection, and derive optimum issue intervals for a certificate revocation list in Public Key Infrastructure.

Entropy was born in information theory [25]. Entropy models were proposed by using its notion and applied to several fields of operations research as useful techniques for solving optimization problems under some constraints [26]. In Sect. 9.3, in contrast with the above direction, we attempt to apply the entropy model to an age replacement policy and other maintenance policies. There exist many problems in other fields that have to be solved by reliability techniques, and inversely, some unnoticed reliability models that can be well adapted to other models.

1.4 Further Studies

Several interesting words as research on new subjects such as system complexity, backward time, traceability, entropy model, and service reliability are presented. Traceability and service reliability will be specially worthy of new topics, and the golden ratio is found twice for the first time in reliability. These terms may not be well defined yet and are roughly discussed. However, they will offer new subjects for further study to researchers and will be needed certainly in practical fields of actual reliability and maintenance models.

Furthermore, several methods and techniques are used in analyzing mathematically proposed models: The partition method for a finite object, the method of obtaining optimum policies sequentially for a finite span, and the methods of solving optimization problems with two variables and of going back from the present time after failure detection will be deeply studied and widely applied to more complex systems. There are many optimization problems existing in other fields of reliability to prevent serious events and to maximize or minimize appropriate objective functions as much as one can. Several examples of showing how to apply reliability techniques to computer and communication systems and to management models will give a good guide to the study of such fields.

Reliability theory has originated from making good products with high quality and designing highly reliable systems. Introducing the concept of safety and risk, reliability has been developed greatly and has been applied widely. We want to make and sell articles with high reliability and buy and use them with safety and no risk in daily life. In addition, we request and hope heartily to use such things with and live in a sense of security, safety, and assurance. These words are combined in one word *Anshin* in Japan whose symbols are filled with towns. We wish to take a new look at reliability theory from the viewpoint of Anshin and establish a new *Anshin Science* in the fields of reliability and maintenance referring to environmental problems of the earth.

2

Redundant Models

It would be necessary to incorporate redundancy into the design of systems to meet the demand for high reliability. We discuss analytically the number of units of redundant systems and their maintenance times mainly based our original work. As some examples of applications, we present typical redundant models in various fields and analyze them from reliability viewpoints. From such results, we can learn practically how to make the design of redundancy and when to do some maintenance. It would be useful for us to acquire redundant techniques in practical situations in other fields.

High system reliability can be achieved by redundancy and maintenance. The most typical model is a standard parallel system that consists of n units in parallel. It was shown by graph that the system can operate for a specified mean time by either changing the replacement time or increasing the number of units [2]. The reliabilities of many redundant systems were computed and summarized [5]. A variety of redundant systems with multiple failure modes and their optimization problems were discussed in detail [27]. Reliabilities of parallel and parallel-series systems with dependent failures of components were derived [28].

First, we summarize our research results for a parallel system with n units in Sect. 2.1 [29, 30]: The optimum number of units and times of two replacement policies are derived analytically. These results are easily extended to a k-out-of-n system [31]. Furthermore, we consider two replacement models where the system is replaced at periodic times if the total number of failed units exceeds a threshold level [29].

Next, in Sect. 2.2, we take up series-parallel and parallel-series systems and analyze theoretically the stochastic behavior of two systems with the same number of series and parallel units. An optimum number of units for a series-parallel system with complexity ([32], see Chap. 9) is also derived.

As one example of analyzing redundant systems, we consider three redundant systems in Sect. 2.3; (1) a one-unit system with n-hold mean time, (2) an n-unit parallel system, and (3) an n-unit standby system. Various kinds of reliability measures of the three systems are computed and compared.

The notion and techniques of redundancy are indispensable in a communication system [12]. Some data transmission models and their optimum schemes were formulated and discussed analytically and numerically [33]. Section 2.4 adopts three schemes of ARQ (Automatic Repeat Request) as the data transmission and discusses which model is the best among the three schemes.

Many stochastic redundant models exist in the general public. Finally, as practical applications, we give three redundant models in Sect. 2.5 [30]; (1) transmission with redundant bits, (2) redundant networks, and (3) redundant copies. The optimum designs of redundancy for the three models are discussed analytically. Redundant techniques of computer systems for improving reliability and achieving fault tolerance have been classified in Sect. 1.2.

2.1 Parallel Systems

System reliabilities can be improved by redundant units. This section summarizes the known results for parallel redundant systems [29,30,34]: First, we derive an optimum number of units for a parallel system with n units. It is shown that similar discussions can be had about a k-out-of-n system. Next, we discuss two replacement policies where the system is replaced at time T. Furthermore, we take up two replacement policies where the system is replaced at periodic times if the total number of failed units exceeds a threshold level.

2.1.1 Number of Units and Replacement Time

(1) Number of Units

Consider a parallel redundant system that consists of n identical units and fails when all units have failed, $i.e.$, when at least one of n units is operating, the system is also operating. Each unit has an independent and identical failure distribution $F(t)$ with a finite mean μ_1. It is assumed that the failure rate is $h(t) \equiv f(t)/\overline{F}(t)$, where $\overline{F}(t) \equiv 1 - F(t)$ and $f(t)$ is a density function of $F(t)$ $i.e.$, $f(t) \equiv dF(t)/dt$. Because the system with n units has a failure distribution $F(t)^n$, its mean time to failure is

$$\mu_n = \int_0^\infty [1 - F(t)^n]\,dt \qquad (n = 1, 2, \ldots), \qquad (2.1)$$

that increases strictly with n from μ_1 to ∞. Therefore, the expected cost rate is [34, 35]

$$C(n) = \frac{nc_1 + c_R}{\mu_n} \qquad (n = 1, 2, \ldots), \qquad (2.2)$$

where c_1 = acquisition cost for one unit and c_R = replacement cost for a failed system.

We find an optimum number n^* that minimizes $C(n)$. Forming the inequality $C(n + 1) - C(n) \geq 0$,

$$\frac{\mu_n}{\mu_{n+1} - \mu_n} - n \geq \frac{c_R}{c_1} \qquad (n = 1, 2, \dots), \qquad (2.3)$$

whose left-hand side increases strictly to ∞ because

$$\frac{\mu_n}{\mu_{n+1} - \mu_n} - n \geq \frac{\mu_1}{\mu_{n+1} - \mu_n} - 1 \qquad \text{for } n \geq 1.$$

Thus, there exists a finite and unique minimum n^* $(1 \leq n^* < \infty)$ that satisfies (2.3) because $\mu_{n+1} - \mu_n$ goes to zero as $n \to \infty$.

In the particular case of $F(t) = 1 - e^{-\lambda t}$,

$$\mu_n = \int_0^\infty [1 - (1 - e^{-\lambda t})^n]\, dt = \frac{1}{\lambda} \sum_{j=1}^n \frac{1}{j}, \qquad (2.4)$$

that is given approximately by

$$\mu_n \approx \frac{1}{\lambda}(C + \log n) \qquad \text{for large } n,$$

where C is Euler's constant and $C = 0.577215\dots$. It was also shown [2, p. 65] that when $F(t)$ is IFR (Increasing Failure Rate), $i.e.$, $h(t)$ increases, a parallel system has a IFR property and

$$\mu_1 \leq \mu_n \leq \mu_1 \sum_{j=1}^n \frac{1}{j}.$$

In addition, (2.3) becomes

$$(n+1) \sum_{j=1}^n \frac{1}{j} - n = \sum_{j=1}^n \frac{n+1}{j+1} \geq \frac{c_R}{c_1} \qquad (n = 1, 2, \dots), \qquad (2.5)$$

whose left-hand side increases strictly from 1 to ∞. Note that an optimum number n^* does not depend on the mean failure time μ_1 of each unit. Because

$$\sum_{j=1}^n \frac{n+1}{j+1} - n = \sum_{j=1}^n \frac{n-j}{j+1} \geq 0,$$

if $n - 1 < c_R/c_1 \leq n$, then $n^* \leq n$. Conversely, because

$$\sum_{j=1}^n \frac{n+1}{j+1} \leq \sum_{j=1}^n j = \frac{n(n+1)}{2},$$

if $\sum_{j=1}^n j < c_R/c_1 \leq \sum_{j=1}^{n+1} j$, then $n^* \geq n$.

(2) Replacement Times

Suppose that a parallel system is replaced at time T $(0 < T \leq \infty)$ or at failure, whichever occurs first. Then, the mean time to replacement is

$$\mu_n(T) \equiv \int_0^T [1 - F(t)^n]\, dt, \tag{2.6}$$

where note that $\mu_n(\infty) = \mu_n$ in (2.1). Thus, the expected cost rate is [29]

$$C_1(T) = \frac{nc_1 + c_R F(T)^n}{\mu_n(T)}. \tag{2.7}$$

When $n = 1$, $C_1(T)$ agrees with the expected cost rate for the standard age replacement [1, p. 72].

We find an optimum replacement time T_1^* that minimizes $C_1(T)$ for a given n $(n \geq 2)$. It is assumed that the failure rate $h(t)$ increases. Then, differentiating $C_1(T)$ with respect to T and setting it equal to zero,

$$H(T)\mu_n(T) - F(T)^n = \frac{nc_1}{c_R}, \tag{2.8}$$

where

$$H(t) \equiv \frac{nh(t)[F(t)^{n-1} - F(t)^n]}{1 - F(t)^n}.$$

It is easily proved that

$$\frac{1 - F(t)^{n-1}}{1 - F(t)^n} = \frac{\sum_{j=0}^{n-2} F(t)^j}{\sum_{j=0}^{n-1} F(t)^j}$$

decreases strictly with t from 1 to $(n-1)/n$ for $n \geq 2$, and

$$\lim_{t \to \infty} \frac{n[F(t)^{n-1} - F(t)^n]}{1 - F(t)^n} = 1.$$

Thus, $H(t)$ increases strictly with t to $h(\infty)$ for $n \geq 2$. Therefore, denoting the left-hand side of (2.8) by $Q_1(T)$, it follows that $\lim_{T \to 0} Q_1(T) = 0$,

$$\frac{dQ_1(T)}{dT} = H'(T)\mu_n(T) > 0, \qquad \lim_{T \to \infty} Q_1(T) = \mu_n h(\infty) - 1,$$

where μ_n is given in (2.1).

Therefore, we have the following optimum policy:

(i) If $\mu_n h(\infty) > (nc_1 + c_R)/c_R$, then there exists a finite and unique T_1^* $(0 < T_1^* < \infty)$ that satisfies (2.8), and the resulting cost rate is

$$C_1(T_1^*) = c_R H(T_1^*). \tag{2.9}$$

(ii) If $\mu_n h(\infty) \leq (nc_1 + c_R)/c_R$, then $T_1^* = \infty$, i.e., the system is replaced only at failure, and the expected cost rate $C_1(\infty)$ is given in (2.2).

Next, suppose that a parallel system is replaced only at time T, i.e., the system remains in a failed state for the time interval from a system failure to its detection at time T. Then, the expected cost rate is [29]

$$C_2(T) = \frac{nc_1 + c_D \int_0^T F(t)^n \, dt}{T},$$
(2.10)

where c_D = downtime cost per unit of time from system failure to replacement. When $n = 1$, $C_2(T)$ agrees with the expected cost rate for the model with no replacement at failure [1, p. 120]. Differentiating $C_2(T)$ with respect to T and setting it equal to zero,

$$\int_0^T [F(T)^n - F(t)^n] \, dt = \frac{nc_1}{c_D}.$$
(2.11)

The left-hand side of (2.11) increases strictly from 0 to μ_n. Therefore, the optimum policy is as follows:

(iii) If $\mu_n > nc_1/c_D$, then there exists a finite and unique T_2^* $(0 < T_2^* < \infty)$ that satisfies (2.11), and the resulting cost rate is

$$C_2(T_2^*) = c_D F(T_2^*)^n.$$
(2.12)

(iv) If $\mu_n \leq nc_1/c_D$, then $T_2^* = \infty$.

Example 2.1. Suppose that the failure time of each unit is exponential, i.e., $F(t) = 1 - e^{-\lambda t}$. Then, we have the respective optimum replacement times T_1^* and T_2^* that minimize $C_1(T)$ in (2.7) and $C_2(T)$ in (2.10) as follows: From the optimum policies (i) and (ii), if $\sum_{j=2}^n 1/j > nc_1/c_R$ for $n \geq 2$, then T_1^* is given by a unique solution of the equation

$$\frac{ne^{-\lambda T}(1 - e^{-\lambda T})^{n-1}}{1 - (1 - e^{-\lambda T})^n} \sum_{j=1}^n \frac{1}{j}(1 - e^{-\lambda T})^j - (1 - e^{-\lambda T})^n = \frac{nc_1}{c_R},$$
(2.13)

and the resulting cost rate is

$$C_1(T_1^*) = \frac{c_R n \lambda e^{-\lambda T_1^*}(1 - e^{-\lambda T_1^*})^{n-1}}{1 - (1 - e^{-\lambda T_1^*})^n}.$$
(2.14)

From (iii) and (iv), if $\sum_{j=1}^n 1/j > n\lambda c_1/c_D$, then T_2^* is a unique solution of the equation

$$\frac{1}{\lambda} \sum_{j=1}^n \frac{1}{j}(1 - e^{-\lambda T})^j - T[1 - (1 - e^{-\lambda T})^n] = \frac{nc_1}{c_D},$$
(2.15)

and the resulting cost rate is

$$C_2(T_2^*) = c_D(1 - e^{-\lambda T_2^*})^n. \tag{2.16}$$

Tables 2.1 and 2.2 present the optimum times T_1^* and T_2^* for c_R/c_1 and c_1/c_D when $1/\lambda = 50$ and $n = 2, 3, 5, 15$, and 20 (see Example 2.2). ∎

(3) k-out-of-n System

Suppose that the system consists of a k-out-of-n system $(1 \le k \le n)$, i.e., it is operating if and only if at least k units of n units are operating [2]. The reliability characteristics of such a system were investigated [36,37]. The number of units that should be on-line to assure that a minimum of k units will be available to complete an assignment for mass transit and computer systems was determined [38]. A k-out-of-n code is also used as a totally self-checking checker for error detecting codes [12]. A good survey of multi-state and consecutive k-out-of-n systems was done [31,39].

The mean time to system failure is [2]

$$\mu_{n,k} = \sum_{j=k}^{n} \binom{n}{j} \int_0^\infty [\overline{F}(t)]^j [F(t)]^{n-j} \, dt, \tag{2.17}$$

and the expected cost rate is

$$C(n; k) = \frac{nc_1 + c_R}{\mu_{n,k}} \qquad (n = k, k+1, \dots). \tag{2.18}$$

When $F(t) = 1 - e^{-\lambda t}$, the expected cost rate is simplified as

$$C(n; k) = \frac{nc_1 + c_R}{(1/\lambda) \sum_{j=k}^{n} (1/j)}, \tag{2.19}$$

and an optimum number that minimizes $C(n, k)$ is obtained by a finite and unique minimum n^* $(k \le n^* < \infty)$ such that

$$(n+1) \sum_{j=k}^{n} \frac{1}{j} - n \ge \frac{c_R}{c_1} \qquad (n = k, k+1, \dots). \tag{2.20}$$

It is natural that n^* increases with k.

Similarly, the expected cost rates in (2.7) and (2.10) are easily written as, respectively,

$$C_1(T, k) = \frac{nc_1 + c_R \sum_{j=0}^{k-1} \binom{n}{j} [\overline{F}(T)]^j [F(T)]^{n-j}}{\sum_{j=k}^{n} \binom{n}{j} \int_0^T [\overline{F}(t)]^j [F(t)]^{n-j} dt}, \tag{2.21}$$

$$C_2(T, k) = \frac{nc_1 + c_D \sum_{j=0}^{k-1} \binom{n}{j} \int_0^T [\overline{F}(t)]^j [F(t)^{n-j}] dt}{T}, \tag{2.22}$$

where all results agree with those of **(1)** and **(2)** when $k = 1$.

In particular, when $k = n - 1$, an $(n - 1)$-out-of-n $(n \geq 3)$ system can be identified with a fault tolerant system with a single bit error correction and referred to a fail-safe design in reliability theory [2, p. 216]. In addition, when $F(t) = 1 - \mathrm{e}^{-\lambda t}$, the expected cost rate in (2.21) is rewritten as

$$C_1(T) = \frac{nc_1 + c_R[1 - n\mathrm{e}^{-(n-1)\lambda T} + (n-1)\mathrm{e}^{-n\lambda T}]}{(1/\lambda)\left\{[n/(n-1)][1 - \mathrm{e}^{-(n-1)\lambda T}] - [(n-1)/n][1 - \mathrm{e}^{-n\lambda T}]\right\}}. \tag{2.23}$$

Differentiating $C_1(T)$ with respect to T and setting it equal to zero,

$$\frac{n(1 - \mathrm{e}^{-\lambda T}) - (1 - \mathrm{e}^{-n\lambda T})}{(n-1)(1 - \mathrm{e}^{-\lambda T}) + 1} = \frac{nc_1}{c_R}. \tag{2.24}$$

The left-hand side of (2.24) increases strictly from 0 to $(n-1)/n$. Thus, if $(n-1)/n > nc_1/c_R$, then there exists a finite and unique T_1^* $(0 < T_1^* < \infty)$ that satisfies (2.24), and the resulting cost rate is

$$C_1(T_1^*) = \frac{c_R \lambda n(n-1)(1 - \mathrm{e}^{-\lambda T_1^*})}{(n-1)(1 - \mathrm{e}^{-\lambda T_1^*}) + 1}. \tag{2.25}$$

Similarly, the expected cost rate in (2.22) is

$$C_2(T) = \frac{nc_1 - (c_D/\lambda)\left\{[n/(n-1)][1 - \mathrm{e}^{-(n-1)\lambda T}] - [(n-1)/n][1 - \mathrm{e}^{-n\lambda T}]\right\}}{T}$$

$$+ c_D. \tag{2.26}$$

Differentiating $C_2(T)$ with respect to T and setting it equal to zero,

$$\frac{n}{n-1}\left[1 - \mathrm{e}^{-(n-1)\lambda T}\right] - \frac{n-1}{n}\left[1 - \mathrm{e}^{-n\lambda T}\right] - \lambda T\left[n\mathrm{e}^{-(n-1)\lambda T} - (n-1)\mathrm{e}^{-n\lambda T}\right]$$

$$= \frac{n\lambda c_1}{c_D}. \tag{2.27}$$

The left-hand side of (2.27) increases strictly from 0 to $1/n + 1/(n-1)$. Thus, if

$$\frac{1}{n} + \frac{1}{n-1} > \frac{n\lambda c_1}{c_D},$$

then there exists a finite and unique T_2^* that satisfies (2.27), and the resulting cost rate is

$$C_2(T_2^*) = c_D[1 - n\mathrm{e}^{-(n-1)\lambda T_2^*} + (n-1)\mathrm{e}^{-n\lambda T_2^*}]. \tag{2.28}$$

2.1.2 Replacement Number of Failed Units

Suppose that the replacement may be done at planned time jT $(j = 1, 2, \dots)$, where T means a day, a week, a month, and so on. Similar preventive maintenance models were considered [1, p. 54, 40]. If the total number of failed units in a parallel system with n units exceeds N $(1 \le N \le n - 1)$ until time $(j + 1)T$, then the system is replaced before failure at time $(j + 1)T$ $(j = 0, 1, 2, \dots)$.

The system is replaced at failure or at time $(j+1)T$ when the total number of failed units has exceeded N, whichever occurs first. Then, the probability that the system is replaced at failure is [29]

$$\sum_{j=0}^{\infty} \sum_{i=0}^{N-1} \binom{n}{i} [F(jT)]^i [F((j+1)T) - F(jT)]^{n-i}, \qquad (2.29)$$

and the probability that it is replaced before failure, *i.e.*, when $N, N + 1, \dots, n - 1$ units have failed until $(j+1)T$, is

$$\sum_{j=0}^{\infty} \sum_{i=0}^{N-1} \binom{n}{i} [F(jT)]^i \sum_{k=N-i}^{n-i-1} \binom{n-i}{k} [F((j+1)T) - F(jT)]^k [\overline{F}((j+1)T)]^{n-i-k},$$

$$(2.30)$$

where note that $(2.29) + (2.30) = 1$. Thus, the mean time to replacement is

$$\sum_{j=0}^{\infty} \sum_{i=0}^{N-1} \binom{n}{i} [F(jT)]^i \int_{jT}^{(j+1)T} t \, \mathrm{d}\{ [F(t) - F(jT)]^{n-i} \}$$

$$+ \sum_{j=0}^{\infty} [(j+1)T] \sum_{i=0}^{N-1} \binom{n}{i} [F(jT)]^i$$

$$\times \sum_{k=N-i}^{n-i-1} \binom{n-i}{k} [F((j+1)T) - F(jT)]^k [\overline{F}((j+1)T)]^{n-i-k}$$

$$= \sum_{j=0}^{\infty} \sum_{i=0}^{N-1} \binom{n}{i} [F(jT)]^i \int_{jT}^{(j+1)T} \{ [\overline{F}(jT)]^{n-i} - [F(t) - F(jT)]^{n-i} \} \, \mathrm{d}t.$$

$$(2.31)$$

Therefore, the expected cost rate is, from (2.7),

$$C_1(N) = \frac{nc_1 + c_R \sum_{j=0}^{\infty} \sum_{i=0}^{N-1} \binom{n}{i} [F(jT)]^i [F((j+1)T) - F(jT)]^{n-i}}{\sum_{j=0}^{\infty} \sum_{i=0}^{N-1} \binom{n}{i} [F(jT)]^i \int_{jT}^{(j+1)T} \{ [\overline{F}(jT)]^{n-i} - [F(t) - F(jT)]^{n-i} \} \, \mathrm{d}t}$$

$$(N = 1, 2, \dots, n). \qquad (2.32)$$

When $N = n$, the system is replaced only at failure, and $C_1(n)$ agrees with $C(n)$ in (2.2).

Next, suppose that the system is replaced only at time $(j + 1)T$ $(j = 0, 1, 2, \dots)$ when the total number of failed units has exceeded N until time $(j + 1)T$. Then, the mean time from system failure to replacement is

$$\sum_{j=0}^{\infty} \sum_{i=0}^{N-1} \binom{n}{i} [F(jT)]^i \int_{jT}^{(j+1)T} [(j+1)T - t]\, \mathrm{d}\{[F(t) - F(jT)]^{n-i}\}$$

$$= \sum_{j=0}^{\infty} \sum_{i=0}^{N-1} \binom{n}{i} [F(jT)]^i \int_{jT}^{(j+1)T} [F(t) - F(jT)]^{n-i}\, \mathrm{d}t, \qquad (2.33)$$

and the mean time to replacement is

$$\sum_{j=0}^{\infty} [(j+1)T] \sum_{i=0}^{N-1} \binom{n}{i} [F(jT)]^i$$

$$\times \sum_{k=N-i}^{n-i} \binom{n-i}{k} [F((j+1)T) - F(jT)]^k [\overline{F}((j+1)T)]^{n-i-k}$$

$$= T \sum_{j=0}^{\infty} \sum_{i=0}^{N-1} \binom{n}{i} [F(jT)]^i [\overline{F}(jT)]^{n-i}, \qquad (2.34)$$

where note that $(2.31) + (2.33) = (2.34)$.

Therefore, the expected cost rate is, from (2.10),

$$C_2(N) = \frac{nc_1 + c_D \sum_{j=0}^{\infty} \sum_{i=0}^{N-1} \binom{n}{i} [F(jT)]^i \int_{jT}^{(j+1)T} [F(t) - F(jT)]^{n-i}\mathrm{d}t}{T \sum_{j=0}^{\infty} \sum_{i=0}^{N-1} \binom{n}{i} [F(jT)]^i [\overline{F}(jT)]^{n-i}}$$

$$(N = 1, 2, \ldots, n). \qquad (2.35)$$

Example 2.2. We compute the respective optimum numbers N_1^* and N_2^* $(1 \le N_i^* \le n)$ that minimize $C_1(N)$ and $C_2(N)$ for a fixed $T > 0$ when $F(t) = 1 - \mathrm{e}^{-\lambda t}$. In this case, the expected cost rate in (2.32) is rewritten as

$$C_1(N) = \frac{nc_1 + c_R \sum_{j=0}^{\infty} \sum_{i=0}^{N-1} \binom{n}{i} (1 - \mathrm{e}^{-j\lambda T})^i [\mathrm{e}^{-j\lambda T} - \mathrm{e}^{-(j+1)\lambda T}]^{n-i}}{(1/\lambda) \sum_{j=0}^{\infty} \sum_{i=0}^{N-1} \binom{n}{i} (1 - \mathrm{e}^{-j\lambda T})^i (\mathrm{e}^{-j\lambda T})^{n-i} \sum_{k=1}^{n-i} [(1 - \mathrm{e}^{-\lambda T})^k / k]}$$

$$(N = 1, 2, \ldots, n). \qquad (2.36)$$

Forming the inequality $C_1(N+1) - C_1(N) \ge 0$,

$$L_1(N) \ge \frac{nc_1}{c_R} \qquad (N = 1, 2, \ldots, n-1), \qquad (2.37)$$

where

$$L_1(N) \equiv \frac{(1 - \mathrm{e}^{-\lambda T})^{n-N}}{\sum_{k=1}^{n-N} [(1 - \mathrm{e}^{-\lambda T})^k / k]}$$

$$\times \sum_{j=0}^{\infty} \sum_{i=0}^{N-1} \binom{n}{i} (1 - \mathrm{e}^{-j\lambda T})^i (\mathrm{e}^{-j\lambda T})^{n-i} \sum_{k=1}^{n-i} \frac{(1 - \mathrm{e}^{-\lambda T})^k}{k}$$

$$- \sum_{j=0}^{\infty} \sum_{i=0}^{N-1} \binom{n}{i} (1 - \mathrm{e}^{-j\lambda T})^i [\mathrm{e}^{-j\lambda T} - \mathrm{e}^{-(j+1)\lambda T}]^{n-i}.$$

Because

$$L_1(N+1) - L_1(N) = \sum_{j=0}^{\infty} \sum_{i=0}^{N} \binom{n}{i}(1 - e^{-j\lambda T})^i (e^{-j\lambda T})^{n-i} \sum_{k=1}^{n-i} \frac{(1 - e^{-\lambda T})^k}{k}$$

$$\times \left\{ \frac{(1 - e^{-\lambda T})^{n-N-1}}{\sum_{k=1}^{n-N-1}[(1 - e^{-\lambda T})^k/k]} - \frac{(1 - e^{-\lambda T})^{n-N}}{\sum_{k=1}^{n-N}[(1 - e^{-\lambda T})^k/k]} \right\}$$

$$> 0,$$

$L_1(N)$ increases strictly with N. Thus, if $L_1(n-1) \geq nc_1/c_R$, then there exists a unique minimum N_1^* ($1 \leq N_1^* \leq n-1$) that satisfies (2.37), and otherwise, $N_1^* = n$, i.e., the system is replaced only at failure.

The expected cost rate $C_2(N)$ in (2.35) is rewritten as

$$C_2(N) = \frac{nc_1 + c_D \sum_{j=0}^{\infty} \sum_{i=0}^{N-1} \binom{n}{i}(1 - e^{-j\lambda T})^i (e^{-j\lambda T})^{n-i}}{T \sum_{j=0}^{\infty} \sum_{i=0}^{N-1} \binom{n}{i}(1 - e^{-j\lambda T})^i (e^{-j\lambda T})^{n-i}}$$

$$(N = 1, 2, \ldots, n). \qquad (2.38)$$

From the inequality $C_2(N+1) - C_2(N) \geq 0$,

$$L_2(N) \geq \frac{n\lambda c_1}{c_D} \qquad (N = 1, 2, \ldots, n-1), \qquad (2.39)$$

where

$$L_2(N) \equiv \sum_{j=0}^{\infty} \sum_{i=0}^{N-1} \binom{n}{i}(1 - e^{-j\lambda T})^i (e^{-j\lambda T})^{n-i} \sum_{k=n-N+1}^{n-i} \frac{(1 - e^{-\lambda T})^k}{k}.$$

It can be easily seen that $L_2(N)$ increases strictly with N. Thus, if $L_2(n-1) \geq n\lambda c_1/c_D$, then there exists a unique minimum N_2^* ($1 \leq N_2^* \leq n-1$) that satisfies (2.39), and otherwise, $N_2^* = n$, i.e., the system is replaced only after failure.

Tables 2.1 and 2.2 present the optimum numbers N_1^* and N_2^* for c_R/c_1 and c_1/c_D when $1/\lambda = 50, T = 4$, and $n = 2, 3, 5, 15$, and 20. For example, when $n = 5$ and $c_R/c_1 = 10$, the mean failure time of the system is $\mu_5 = 114.2$, and the optimum time is $T_1^* = 75.0$, i.e., the system should be replaced at $(75.0/114.2) \times 100 = 65.4\%$ of its mean time from Table 2.1. Such percentages increase with n and decrease with c_R/c_1. In the same case, $N_1^* = 3$, i.e., the system should be replaced when at least three of five units have failed at some jT. Such optimum numbers also increase with n and decrease with c_R/c_1. We can give a similar explanation in Table 2.2. It is of interest that the system should be replaced when $n-1$ or $n-2$ units have failed in both tables. ∎

Table 2.1. Optimum time T_1^* and number N_1^* when $1/\lambda = 50$, $T = 4$

c_R/c_1	5		10		20		30		40		50	
n	T_1^*	N_1^*	T_1^*	N_1^*	T_1^*	N_1^*	T_1^*	N_1^*	T_1^*	N_1^*	T_1^*	N_1^*
2	99.6	1	40.9	1	23.2	1	16.2	1	14.4	1	12.5	1
3	99.6	1	51.5	1	33.4	1	26.9	1	23.3	1	21.1	1
5	135.3	3	75.0	3	53.0	3	44.9	3	40.4	2	37.4	2
15	∞	13	159.2	13	114.9	13	101.5	13	94.1	13	89.3	12
20	∞	18	197.2	18	135.7	18	119.7	18	111.3	18	105.8	18

Table 2.2. Optimum time T_2^* and number N_2^* when $1/\lambda = 50$, $T = 4$

c_1/c_D	0.05		0.10		0.5		1		5		10	
n	T_2^*	N_2^*	T_2^*	N_2^*	T_2^*	N_2^*	T_2^*	N_2^*	T_2^*	N_2^*	T_2^*	N_2^*
2	107.3	1	67.0	1	30.9	1	23.2	1	12.6	1	9.8	1
3	142.8	2	88.1	2	44.1	2	34.7	2	21.0	1	17.1	1
5	229.0	4	122.7	4	65.4	4	53.6	3	36.0	3	30.8	3
15	∞	15	287.1	14	127.1	14	109.0	14	82.9	13	75.1	13
20	∞	20	∞	19	146.6	19	126.2	19	97.6	18	89.2	18

2.2 Series and Parallel Systems

System reliabilities can be improved by redundant compositions of units. An optimum number of subsystems for a parallel-series system was obtained by considering two failures of open-circuits and short-circuits [2]. Reliability optimization of parallel-series and series-parallel systems was discussed [27]. It has been well-known that the reliability of series-parallel system with n subsystems in series, each subsystem having m units in parallel (Fig. 2.1), goes to 1 as $m \to \infty$ and to 0 as $n \to \infty$. Of interest is the question of what the stochastic behavior of such a system with the same number of series and parallel units is, *i.e.*, $n = m$, as $n \to \infty$. We answer this question mathematically and investigate several characteristics of series-parallel and parallel-series systems.

2.2.1 Series-parallel System

We consider a series-parallel system that consists of n ($n \geq 1$) subsystems in series, each subsystem having identical m ($m \geq 1$) units in parallel(Fig. 2.1). It is assumed that each unit has an identical and independent reliability $q \equiv 1-p$ ($0 < q \leq 1$). Then, the system reliability is [2]

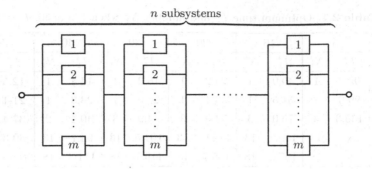

Fig. 2.1. Series-parallel system

$$R_{n,m}(q) = [1 - (1 - q)^m]^n = (1 - p^m)^n \qquad (n, m = 1, 2, \dots). \qquad (2.40)$$

We investigate the characteristics of $R_{n,m}(q)$:

(1) $R_{n,m}(q)$ is an increasing function of q from 0 to 1 because

$$\lim_{q \to 0} R_{n,m}(q) = 0, \qquad \lim_{q \to 1} R_{n,m}(q) = 1.$$

(2) For a fixed p $(0 < p < 1)$, $m < \infty$, and $n < \infty$,

$$\lim_{n \to \infty} R_{n,m}(q) = 0, \qquad \lim_{m \to \infty} R_{n,m}(q) = 1. \qquad (2.41)$$

(3) Using the binomial expansion in (2.40) for $n \geq 2$ and a fixed p $(0 < p < 1)$,

$$1 - np^m < (1 - p^m)^n < 1 - np^m + \frac{n(n-1)}{2}p^{2m}, \qquad (2.42)$$

and hence,

$$0 < R_{n,m}(q) - (1 - np^m) < \frac{n(n-1)}{2}p^{2m}. \qquad (2.43)$$

When $n = m$, we investigate the characteristics of reliability

$$R_n(p) \equiv (1 - p^n)^n \qquad (n = 1, 2, \dots). \qquad (2.44)$$

(4) If n increases to $n + 1$, then the number of units increases in order of n^2. From (2.42), for a fixed p $(0 < p < 1)$,

$$\lim_{n \to \infty} R_n(p) \geq \lim_{n \to \infty} (1 - np^n) = 1.$$

Thus,

$$\lim_{n \to \infty} R_n(p) = 1. \qquad (2.45)$$

Table 2.3. Values of $(1 - p^n)^n$ and $1 - np^n$ when $p = 0.1$, and λMTTF when $p = 1 - e^{-\lambda t}$

n	$(1 - p^n)^n$	$1 - np^n$	λMTTF
1	0.90000	0.90000	1.00000
2	0.98010	0.98000	0.91667
3	0.99700	0.99700	0.97897
4	0.99960	0.99960	1.05830
5	0.99995	0.99995	1.13653

Example 2.3.   We compute the reliability $(1 - p^n)^n$ and its approximation $1 - np^n$ in Table 2.3 for $n = 1$, 2, 3, 4, and 5 when $p = 0.1$. The accuracy of approximation becomes better as p is smaller. In general, it would be sufficient to compute $1 - np^m$ for a series-parallel system for a small p.

In particular, when $p = 1 - e^{-\lambda t}$, the MTTF (Mean Time to Failure) of the system is

$$\int_0^\infty [1 - (1 - e^{-\lambda t})^n]^n \, dt = \frac{1}{\lambda} \int_0^1 \frac{(1 - x^n)^n}{1 - x} \, dx. \qquad (2.46)$$

In particular, when $n = 2$, λMTTF is $11/12$. Table 2.3 also presents λMTTF. It is of great interest that these values are minimum at $n = 2$ and increase strictly with n ($n \geq 2$). ∎

(5) To investigate the monotonic property $R_n(p)$ ($0 < p < 1$) in (2.44) for n, we obtain a solution p_n that satisfies

$$(1 - p^{n+1})^{n+1} = (1 - p^n)^n \qquad (n = 1, 2, \dots). \qquad (2.47)$$

In the particular case of $n = 1$, $p_1 = (-1 + \sqrt{5})/2 \approx 0.618$ that is equal to the golden ratio and plays a role in analyzing the system. We expect that p_n would increase with n. In the approximation $1 - np^n$, a solution to satisfy

$$1 - (n + 1)p^{n+1} = 1 - np^n \qquad (2.48)$$

is given by $\widetilde{p}_n = n/(n + 1)$, that increases from $1/2$ to 1. Thus, $1 - np^n$ increases with n from q to 1 for $0 < p < 1/2$.

Example 2.4.   Table 2.4 presents the values p_n and $\widetilde{p}_n$ for $n = 1$, 2, 3, 4, 5, and 10. This indicates that p_n increases with n and $p_n > \widetilde{p}_n$. Thus, if $0 < p < p_1$, then $R_n(p)$ increases with n. In general, because p is a failure probability, its value would be lower than p_1. Therefore, it might be said in actual fields that $R_n(p)$ could be regarded as an increasing function of n.

It has been well-known that $R_n(p)$ has an S-shape [2, p. 198]. Figure 2.2 draws the reliability $R_n(p)$ for p when $n = 1$, 2, 3, and 4. Because

Table 2.4. Values of p_n, $\tilde{p}_n$, $\hat{p}_n$, and $[1/(n+1)]^{1/n}$

n	p_n	$\tilde{p}_n$	$\hat{p}_n$	$[1/(n+1)]^{1/n}$
1	0.61803	0.50000	0.61803	0.50000
2	0.75488	0.66667	0.68233	0.57735
3	0.81917	0.75000	0.72449	0.62996
4	0.85668	0.80000	0.75809	0.66874
5	0.88127	0.83333	0.77809	0.69883
10	0.93607	0.90909	0.84440	0.78963

$$\frac{\mathrm{d}^2 R_n(p)}{\mathrm{d}p^2} = -(n-1)n^2(p-p^{n+1})^{n-2}[1-(n+1)p^n],$$

the inflection point is $[1/(n+1)]^{1/n}$ for $0 < p < 1$. These points increase with n $(1 \le n < \infty)$ from 0.5 to 1 because the function $[x/(1+x)]^x$ decreases from 1 to 0.5 for $0 < x \le 1$. This is also obtained by setting the approximation reliability is equal to that of one unit, *i.e.*,

$$1 - (n+1)p^{n+1} = 1 - p.$$

Table 2.4 also presents the value $\hat{p}_n$ of a solution of the equation

$$(1 - p^{n+1})^{n+1} = 1 - p,$$

and the inflection points $[1/(n+1)]^{1/n}$ for $n = 1, 2, 3, 4, 5$, and 10. It is obvious that $p_n > \hat{p}_n > [1/(n+1)]^{1/n}$ for $n \ge 2$. From Table 2.4 and Fig. 2.2, if $p > p_1$ that is the golden ratio, then we should not build up such a redundancy system. For example, when the failure time of each unit is exponential, *i.e.*, $p = 1 - e^{-\lambda t}$, we should work this system in the interval less than $t = -[\log(1 - p_1)]/\lambda \approx 0.9624/\lambda$, that is a little smaller than the mean time $1/\lambda$ of a unit. ∎

2.2.2 Parallel-series System

We consider a parallel-series system that consists of m $(m \ge 1)$ subsystems in parallel, each subsystem having identical n $(n \ge 1)$ units in series (Fig. 2.3). When each unit has an identical reliability q $(0 < q \le 1)$, the system reliability is [2]

$$R_{m,n}(q) = 1 - (1 - q^n)^m \qquad (n, m = 1, 2, \ldots). \tag{2.49}$$

When $q = e^{-\lambda t}$, the MTTF is

$$\int_0^\infty [1 - (1 - e^{-n\lambda t})^m]\mathrm{d}t = \frac{1}{n\lambda} \sum_{j=1}^m \frac{1}{j} \qquad (m = 1, 2, \ldots). \tag{2.50}$$

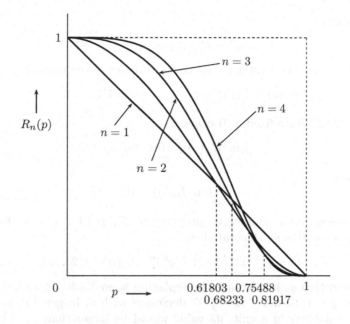

Fig. 2.2. Reliability of series-parallel systems

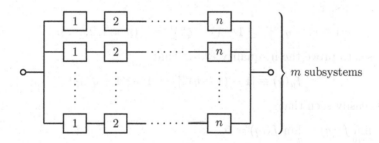

Fig. 2.3. Parallel-series system

When $n = m$, the MTTF decreases with n from $1/\lambda$ to 0.

We investigate the characteristics of $R_{m,n}(q)$:

(1) $R_{m,n}(q)$ increases with q from 0 to 1.

(2) For a fixed q $(0 < q \leq 1)$, $m < \infty$, and $n < \infty$,

$$\lim_{n \to \infty} R_{m,n}(q) = 0, \qquad \lim_{m \to \infty} R_{m,n}(q) = 1. \qquad (2.51)$$

(3) For $m \geq 2$ and a fixed q $(0 < q < 1)$,

$$mq^n - \frac{m(m-1)}{2}q^{2n} < 1 - (1 - q^n)^m < mq^n, \qquad (2.52)$$

and hence,

$$0 < mq^n - R_{m,n}(q) < \frac{m(m-1)}{2}q^{2n}. \tag{2.53}$$

When $n = m$, we investigate the characteristics of the reliability

$$R_n(q) \equiv 1 - (1 - q^n)^n \qquad (n = 1, 2, \dots). \tag{2.54}$$

(4) From (2.52), for a fixed q $(0 < q < 1)$,

$$\lim_{n \to \infty} R_n(q) \le \lim_{n \to \infty} nq^n = 0,$$

and hence

$$\lim_{n \to \infty} R_n(q) = 0. \tag{2.55}$$

(5) To investigate the monotonic property of $R_n(q)$ $(0 < q < 1)$ for n, we obtain a solution q_n that satisfies

$$(1 - q^{n+1})^{n+1} = (1 - q^n)^n \qquad (n = 1, 2, \dots). \tag{2.56}$$

It is clear that q_n is calculated by replacing p_n in Table 2.4 with q_n. Thus, if $p_1 < q < 1$, then $1 - (1 - q^n)^n$ decreases with n. In general, because q is the reliability of a unit, its value would be larger than p_1. Therefore, $R_n(q)$ could be regarded as a decreasing function of n in actual fields.

(6) Comparing two reliabilities of series-parallel and parallel-series systems, for $0 < q < 1$,

$$[1 - (1 - q)^n]^n \ge 1 - (1 - q^n)^n \qquad (n = 1, 2, \dots). \tag{2.57}$$

We set to prove the inequality (2.57) that

$$f_n(q) \equiv [1 - (1 - q)^n]^n - 1 + [(1 - q^n)]^n.$$

It is easily seen that

$$\lim_{q \to 0} f_n(q) = \lim_{q \to 1} f_n(q) = 0,$$

$$f_1(q) = 0, \qquad f_2(q) = 2q^2(1 - q)^2 > 0,$$

$$\frac{\mathrm{d}f_n(q)}{\mathrm{d}q} = n^2 \left\{ [1 - (1 - q)^n]^{n-1}(1 - q)^{n-1} - (1 - q^n)^{n-1}q^{n-1} \right\}$$

$$= n^2 [q(1 - q)]^{n-1} \left\{ \left[\frac{1 - (1 - q)^n}{q} \right]^{n-1} - \left[\frac{1 - q^n}{1 - q} \right]^{n-1} \right\}$$

$$= n^2 [q(1 - q)]^{n-1} \left\{ \left[\sum_{j=0}^{n-1}(1 - q)^j \right]^{n-1} - \left[\sum_{j=0}^{n-1} q^j \right]^{n-1} \right\}.$$

Hence, $\mathrm{d}f_n(q)/\mathrm{d}q > 0$ for $0 < q < 1/2$, 0 for $q = 1/2$, and < 0 for $1/2 < q < 1$. Thus $f_n(q)$ is a concave function of q $(0 < q < 1)$ and takes 0 at $q = 0, 1$. This completes the proof of inequality (2.57). The inequality holds only when $n = 1$, and its difference is maximum at $q = 1/2$.

Table 2.5. Optimum number n^* of series-parallel system with complexity

α	$1 - 10^{-1}$	$1 - 10^{-2}$	$1 - 10^{-3}$
10^{-1}	1	1	1
10^{-2}	2	1	1
10^{-3}	2	2	1
10^{-4}	3	2	2
10^{-5}	4	2	2
10^{-6}	4	3	2
10^{-7}	5	3	2
10^{-8}	5	4	3

2.2.3 Complexity of Series-parallel System

We define the complexity of redundant systems as the number P_a of paths and its reliability as $\exp\{-\alpha[P_a - 1]\}$ that will be denoted in Chap. 9. Based on such definitions, the number of paths of a series-parallel system is $P_a = m^n$, and hence, its reliability is, from (9.6),

$$R_s(n, m) = \exp[-\alpha(m^n - 1)][1 - (1 - q)^m]^n \qquad (n, m = 1, 2, \ldots) \qquad (2.58)$$

for $0 \le \alpha < \infty$. More detailed studies on system complexity will be done in Chap. 9.

Example 2.5. We can obtain the optimum number n^* that maximizes $R_s(n, n)$. The reliability of a series-parallel system increases with n for large q, however, the reliability of the complexity decreases with n. Table 2.5 presents the optimum n^* for $\alpha = 10^{-1}$–10^{-8} and $q = 1 - 10^{-1}, 1 - 10^{-2}$, and $1 - 10^{-3}$. This indicates naturally that n^* decreases with both α and q. ∎

2.3 Three Redundant Systems

As one application of redundant techniques, this section considers the following three typical redundant systems and evaluates them to make the optimization design of system redundancies:

(1) System 1: One unit system with n-fold mean time.
(2) System 2: n-unit parallel redundant system.
(3) System 3: n-unit standby redundant system.

When $n = 1$, all systems are identical. When the failure time of each unit is exponential, we compute the reliability quantities of the three systems.

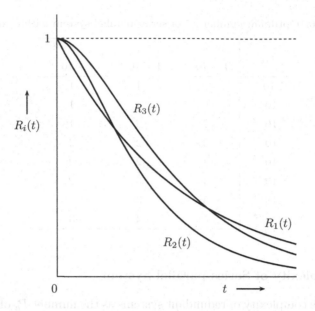

Fig. 2.4. Reliabilities of the three systems when $n = 2$

Furthermore, we obtain the expected costs for each system and compare them. The scheduling problem in which a job has a random working time and is achieved by a system will be discussed in Sect. 5.3. In this model, we define the reliability as the probability that the work of a job is accomplished by a system without failure, derive the reliabilities of the three systems, and compare them.

2.3.1 Reliability Quantities

When the failure time of each unit is exponential, *i.e.*, the failure distribution is $F(t) = 1 - e^{-\lambda t}$, we calculate the following reliability quantities [2]:

(i) Reliability function $R(t)$

(1) $R_1(t) = e^{-\lambda t/n}$,   (2.59)

(2) $R_2(t) = 1 - (1 - e^{-\lambda t})^n$,   (2.60)

(3) $R_3(t) = \displaystyle\sum_{j=0}^{n-1} \frac{(\lambda t)^j}{j!} e^{-\lambda t}$.   (2.61)

Figure 2.4 shows the reliabilities $R_i(t)$ $(i = 1, 2, 3)$ of the three systems when $n = 2$. We can prove that $R_3(t) > R_2(t)$ for $t > 0$ and $n \geq 2$, *i.e.*,

$$(1 - e^{-\lambda t})^n > \sum_{j=n}^{\infty} \frac{(\lambda t)^j}{j!} e^{-\lambda t} \qquad (n = 2, 3, \ldots), \qquad (2.62)$$

by mathematical induction: When $n = 2$, we denote $Q(t)$ by

$$Q(t) \equiv (1 - e^{-\lambda t})^2 - [1 - (1 + \lambda t)e^{-\lambda t}].$$

Then, it is clearly seen that $Q(0) = Q(\infty) = 0$, and

$$\frac{dQ(t)}{dt} = \lambda e^{-\lambda t}[2(1 - e^{-\lambda t}) - \lambda t],$$

that implies $Q(t) > 0$ for $t > 0$ because $Q(t)$ is a concave function. Assuming that when $n = k$,

$$(1 - e^{-\lambda t})^k > \sum_{j=k}^{\infty} \frac{(\lambda t)^j}{j!} e^{-\lambda t},$$

we prove that

$$(1 - e^{-\lambda t})^{k+1} > \sum_{j=k+1}^{\infty} \frac{(\lambda t)^j}{j!} e^{-\lambda t}.$$

We easily have

$$(1 - e^{-\lambda t})^{k+1} - \sum_{j=k+1}^{\infty} \frac{(\lambda t)^j}{j!} e^{-\lambda t}$$

$$> (1 - e^{-\lambda t}) \sum_{j=k}^{\infty} \frac{(\lambda t)^j}{j!} e^{-\lambda t} - \sum_{j=k+1}^{\infty} \frac{(\lambda t)^j}{j!} e^{-\lambda t}$$

$$= e^{-\lambda t} \left[\frac{(\lambda t)^k}{k!} - \sum_{j=k}^{\infty} \frac{(\lambda t)^j}{j!} e^{-\lambda t} \right]$$

$$= \frac{(\lambda t)^k}{k!} e^{-2\lambda t} \left[\sum_{j=0}^{\infty} \frac{(\lambda t)^j}{j!} - \sum_{j=k}^{\infty} \frac{(\lambda t)^j}{j!} \frac{k!}{(\lambda t)^k} \right]$$

$$= \frac{(\lambda t)^k}{k!} e^{-2\lambda t} \sum_{j=0}^{\infty} \frac{(\lambda t)^j}{j!(j+k)!} [(j+k)! - j!k!] > 0.$$

This concludes that $R_3(t) > R_2(t)$ for $n = 2, 3, \ldots$ and $t > 0$.

(ii) Mean time μ and standard deviation σ

(1) $\quad \mu_1 = \dfrac{n}{\lambda}, \qquad\qquad\qquad \sigma_1 = \dfrac{n}{\lambda},$ \hfill (2.63)

(2) $\quad \mu_2 = \dfrac{1}{\lambda} \sum_{j=1}^{n} \dfrac{1}{j}, \qquad\qquad \sigma_2 = \dfrac{1}{\lambda} \sqrt{\sum_{j=1}^{n} \dfrac{1}{j^2}},$ \hfill (2.64)

(3) $\quad \mu_3 = \dfrac{n}{\lambda}, \qquad\qquad\qquad \sigma_3 = \dfrac{\sqrt{n}}{\lambda}.$ \hfill (2.65)

Note that $\mu_1 = \mu_3 > \mu_2$ and $\sigma_1 > \sigma_3 > \sigma_2$ for $n = 2, 3, \ldots$.

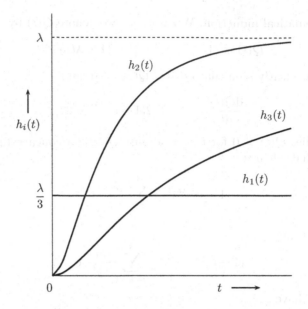

Fig. 2.5. Failure rates of the three systems when $n = 3$

(iii) Failure rate $h(t)$

$$(1) \quad h_1(t) = \frac{\lambda}{n}, \tag{2.66}$$

$$(2) \quad h_2(t) = \frac{n\lambda e^{-\lambda t}(1 - e^{-\lambda t})^{n-1}}{1 - (1 - e^{-\lambda t})^n}, \tag{2.67}$$

$$(3) \quad h_3(t) = \frac{\lambda(\lambda t)^{n-1}/(n-1)!}{\sum_{j=0}^{n-1}[(\lambda t)^j/j!]}. \tag{2.68}$$

Both $h_2(t)$ and $h_3(t)$ increase strictly from 0 to λ for $n \geq 2$. It seems certain that $h_2(t) \geq h_3(t)$. Unfortunately, we cannot prove this inequality mathematically. Figure 2.5 shows the three failure rates $h_i(t)$ when $n = 3$.

(iv) Complexity

When a redundant system has the number n of paths, we define its complexity as $P_e = \log_2 n$ and its reliability as $R_e(n) = \exp(-\alpha \log_2 n)$ for parameter $\alpha > 0$ as shown in Sect. 9.2. Because we count that the numbers of paths are 1 for System 1 and n for Systems 2 and 3, the complexities for System i are $\log_2 1$, $\log_2 n$, and $\log_2 n$, respectively. Thus, the reliabilities of complexity are $\exp(-\alpha \log_2 1)$ for System 1 and $\exp(-\alpha \log_2 n)$ for Systems 2 and 3. If the reliabilities of a whole system with complexity is given by the product of the reliabilities of the system and complexity, then from (2.59)–(2.61),

Table 2.6. Reliabilities $R_i(2)$ (%) of the three systems when $\lambda t = 1$

α	$R_1(2)$	$R_2(2)$	$R_3(2)$
0.2	60.7	49.2	60.2
0.1	60.7	54.3	66.6
0.01	60.7	59.4	72.8
0.001	60.7	60.0	73.5
0.0001	60.7	60.0	73.6

(1) $\quad R_1(n) = e^{-\lambda t/n} \exp(-\alpha \log_2 1) = e^{-\lambda t/n},$ (2.69)

(2) $\quad R_2(n) = [1 - (1 - e^{-\lambda t})^n] \exp(-\alpha \log_2 n),$ (2.70)

(3) $\quad R_3(n) = \sum_{j=0}^{n-1} \frac{(\lambda t)^j}{j!} e^{-\lambda t} \exp(-\alpha \log_2 n).$ (2.71)

Example 2.6. Table 2.6 presents reliabilities $R_i(n)$ of the three systems for $\alpha = 0.2,\ 0.1,\ 0.01,\ 0.001,$ and 0.0001 when $n = 2$ and $\lambda t = 1$. This indicates that System 3 is better than System 2 for any $\alpha > 0$, as shown in (i). When α is larger, System 1 is better than System 3, and when α is smaller, System 3 is better than System 1. When $\alpha = 0.193$, the reliabilities of Systems 1 and 3 are equal to each other. ■

2.3.2 Expected Costs

We introduce the following costs for the three systems:

(1) $\quad C_1(n) = c_1(n) + b + c,$ (2.72)

(2) $\quad C_2(n) = an + bn + c,$ (2.73)

(3) $\quad C_3(n) = an + b + cn,$ (2.74)

where $c_1(n)$ and an are production costs for Systems 1 and 2, 3, where $c_1(1) = a$, b and bn are operating costs for System 1, 3 and 2, and c and cn are replacement costs for System 1, 2 and 3, respectively.

Comparing the three costs,

(i) $\quad c_1(n) \geq an + b(n-1) \iff C_1(n) \geq C_2(n).$ (2.75)

(ii) $\quad c_1(n) \geq an + c(n-1) \iff C_1(n) \geq C_3(n).$ (2.76)

(iii) $\quad b \geq c \iff C_2(n) \geq C_3(n).$ (2.77)

Furthermore, we obtain the following expected cost rates by dividing $C_i(n)$ by the mean times μ_i $(i = 1, 2, 3)$:

(1) $\widetilde{C}_1(n) = \dfrac{c_1(n) + b + c}{n/\lambda}$,   (2.78)

(2) $\widetilde{C}_2(n) = \dfrac{an + bn + c}{(1/\lambda)\sum_{j=1}^{n}(1/j)}$,   (2.79)

(3) $\widetilde{C}_3(n) = \dfrac{an + b + cn}{n/\lambda}$.   (2.80)

Comparing the above three costs,

(iv) $\dfrac{c_1(n) + b + c}{n}\sum_{j=1}^{n}\dfrac{1}{j} \geq an + bn + c \Longleftrightarrow \widetilde{C}_1(n) \geq \widetilde{C}_2(n)$.   (2.81)

(v) $c_1(n) \geq an + c(n-1) \Longleftrightarrow \widetilde{C}_1(n) \geq \widetilde{C}_3(n)$.   (2.82)

(vi) $an + bn + c \geq \dfrac{an + b + cn}{n}\sum_{j=1}^{n}\dfrac{1}{j} \Longleftrightarrow \widetilde{C}_2(n) \geq \widetilde{C}_3(n)$.   (2.83)

Note that the above results do not depend on the failure rate λ of a unit.

(vii) When $c_1(n) = an^2$, we find an optimum number n^* that minimizes

$$\frac{\widetilde{C}_1(n)}{\lambda} = an + \frac{b+c}{n} \qquad (n = 1, 2, \dots).$$ (2.84)

From the inequality $\widetilde{C}_1(n+1) - \widetilde{C}_1(n) \geq 0$,

$$\frac{n(n+1)}{2} \geq \frac{b+c}{2a}.$$ (2.85)

Thus, there exists a finite and unique minimum n^* that satisfies (2.85). Note that the left-hand side represents the summation of integers from 1 to n and will appear often in partition models of Sect. 3.1.

(viii) We find an optimum number n^* to minimize $\widetilde{C}_2(n)$ in (2.79). From the inequality $\widetilde{C}_2(n+1) - \widetilde{C}_2(n) \geq 0$,

$$(n+1)\sum_{j=1}^{n}\frac{1}{j+1} \geq \frac{c}{a+b}.$$ (2.86)

The left-hand side of (2.86) agrees with that of (2.5) and increases strictly to ∞. Thus, there exists a finite and unique minimum n^* that satisfies (2.86).

2.3.3 Reliabilities with Working Time

Suppose that a positive random variable S with distribution $W(t) = \Pr\{S \leq t\}$ is the working time of a job that has to be achieved by each system. Then, we define the reliabilities of each system by

$$R_i \equiv \int_0^\infty R_i(t)\, dW(t) \qquad (i = 1, 2, 3), \tag{2.87}$$

that represent the probabilities that a job with working time S is accomplished by each system without failure. Several properties of these reliabilities and optimization problems are summarized in Sect. 5.3.

From this definition, the reliabilities of the three systems are

$$(1) \quad R_1 = \int_0^\infty e^{-\lambda t/n}\, dW(t), \tag{2.88}$$

$$(2) \quad R_2 = \int_0^\infty [1 - (1 - e^{-\lambda t})^n]\, dW(t), \tag{2.89}$$

$$(3) \quad R_3 = \sum_{j=0}^{n-1} \int_0^\infty \frac{(\lambda t)^j}{j!} e^{-\lambda t}\, dW(t). \tag{2.90}$$

When $W(t) = 1 - e^{-\omega t}$, the above reliabilities are rewritten as

$$(1) \quad R_1 = \frac{n\omega}{\lambda + n\omega}, \tag{2.91}$$

$$(2) \quad R_2 = 1 - \sum_{j=0}^{n} \binom{n}{j}(-1)^j \frac{\omega}{\omega + j\lambda}, \tag{2.92}$$

$$(3) \quad R_3 = 1 - \left(\frac{\lambda}{\lambda + \omega}\right)^n. \tag{2.93}$$

We have the following results:

(i) When $\lambda = \omega$,

$$R_1 = R_2 = \frac{n}{n+1}, \qquad R_3 = 1 - \frac{1}{2^n}, \tag{2.94}$$

and $R_3 > R_1 = R_2$ for $n = 2, 3, \ldots$.

(ii) We compare R_1 and R_3 for $n = 2, 3, \ldots$. Because

$$\frac{\lambda}{\lambda + n\omega} - \left(\frac{\lambda}{\lambda + \omega}\right)^n = \frac{\lambda}{(\lambda + n\omega)(\lambda + \omega)^n}[(\lambda + \omega)^n - \lambda^{n-1}(\lambda + n\omega)]$$

$$= \frac{\lambda}{(\lambda + n\omega)(\lambda + \omega)^n} \sum_{j=0}^{n-2} \binom{n}{j} \lambda^j \omega^{n-j} > 0,$$

$R_3 > R_1$ for $n = 2, 3, \ldots$.

(iii) $R_3 > R_2$ for $n = 2, 3, \ldots$ from $R_3(t) > R_2(t)$.

(iv) When $n = 2, 3$, it is easily proved that

$$\omega > \lambda \iff R_2 > R_1.$$

Furthermore, it seems that this result holds for $n = 4, 5, \ldots$. Unfortunately, we cannot prove it mathematically. Figure 2.6 shows R_1, R_2, and R_3 for $\lambda = a\omega$ $(0 \le a \le 1)$ when $n = 4$.

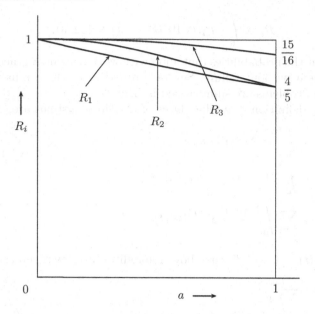

Fig. 2.6. Reliabilities of the three systems when $n = 4$

2.4 Redundant Data Transmissions

Data transmissions in a communication system fail due to some errors that have been generated by disconnection, cutting, warping, noise, or distortion in a communication line. To transmit accurate data, we have to prepare error control schemes that automatically detect and correct errors. The following three control schemes have been used mainly in communication systems [41–43]: (1) FEC (Forward Error Connection) scheme, (2) ARQ (Automatic Repeat Request) scheme, and (3) Hybrid ARQ scheme. A variety of such error-correcting strategies and a great many protocols of ARQ schemes were proposed and appeared in actual systems [33].

Scheme 2 has been widely used in data transmissions between two points because the error control is simple and easy. This section considers three simple models of Scheme 2 and obtains the expected costs until the success of data transmission [44]. We discuss analytically which model is the best among three models. The techniques used in this section would be useful for other schemes.

2.4.1 Three Models

We transmit an amount of data from a sender to a receiver that is called *unit data*. To detect and correct errors, we consider three redundant models, where we transmit two, two plus one, and three unit data simultaneously to a receiver.

Suppose that the transmission of unit data fails with probability p ($0 \leq p < 1$) due to errors that have occurred independently of each other. If there is no failure of the transmission, all transmitted data are the same ones at a receiver. Let c_n ($n = 1, 2, \ldots$) be the cost required for the transmissions of n unit data; this includes all costs of editing, transmission, and checking. It is assumed that $c_2 + c_1 > c_3 > c_2 > c_1$.

(1) Model 1

We transmit two unit data simultaneously to a receiver who checks two data:

(1) If the two data are not the same, then the receiver cancels such data and informs the sender. We call it a transmission failure.
(2) If the two data are the same, the receiver accepts the data and informs the sender. We call it a transmission success.
(3) When the transmission has failed, the sender transmits two units data again and continues the above transmission until its success.

The expected cost until transmission success is given by a renewal equation:

$$C_1 = (1 - p)^2 c_2 + [1 - (1 - p)^2](c_2 + C_1). \qquad (2.95)$$

Solving (2.95) for C_1,

$$C_1 = \frac{c_2}{(1 - p)^2}. \qquad (2.96)$$

(2) Model 2

(1) If the two data are not the same, the sender transmits one unit data again and the receiver checks it with two former data. If the retransmitted data are not the same as either of two data, we call it a transmission failure and transmit the two unit data from the beginning.
(2) If the two data are the same or if the retransmitted data are the same as either of two former data, we call it a transmission success.
(3) The sender continues the above transmission until its success.

The expected cost is

$$C_2 = (1 - p)^2 c_2 + 2p(1 - p)^2 (c_2 + c_1) + [2p^2(1 - p) + p^2](c_2 + c_1 + C_2),$$

i.e.,

$$C_2 = \frac{c_2 + [1 - (1 - p)^2]c_1}{(1 - p)^2(1 + 2p)}. \qquad (2.97)$$

(3) Model 3

We transmit three unit data simultaneously to a receiver who checks them:

(1) If none of the three data are the same, the receiver cancels such data and informs the sender. We call it a transmission failure.
(2) If at least two of the three data are the same, the receiver accepts the data. We call it a transmission success.
(3) The sender continues the above transmission until its success.

The probability that at least two of the three data are the same is

$$\sum_{j=0}^{1} \binom{3}{j} p^j (1-p)^{3-j} = (1-p)^2(1+2p),$$

that agrees with the denominator in (2.97). Thus, the expected cost is

$$C_3 = \frac{c_3}{(1-p)^2(1+2p)}. \tag{2.98}$$

Note that all expected costs increase with p from $C_1 = C_2 = c_2$ and $C_3 = c_3$ to ∞.

If we transmit n $(n \geq 3)$ unit data simultaneously and if at least two of n data are the same, we call it a transmission success. Then, the expected cost is similarly given by

$$C_3 = \frac{c_n}{1 - np^{n-1} + (n-1)p^n}. \tag{2.99}$$

2.4.2 Optimum Policies

We compare the three expected costs C_1, C_2, and C_3. From (2.96) and (2.97),

$$C_1 - C_2 = \frac{1}{(1-p)^2(1+2p)}[2p(c_2 - c_1) + p^2 c_1] > 0,$$

that implies $C_1 > C_2$. In addition, from (2.96) − (2.98),

$$C_3 - C_2 = \frac{1}{(1-p)^2(1+2p)}[(1-p)^2 c_1 - (c_2 + c_1 - c_3)],$$

$$C_1 - C_3 = \frac{1}{(1-p)^2(1+2p)}[2pc_2 - (c_3 - c_2)].$$

Therefore, we have the following optimum policy:

(i) If $(1-p)^2 \leq (c_2 + c_1 - c_3)/c_1$, then $C_1 > C_2 \geq C_3$.
(ii) If $(1-p)^2 > (c_2 + c_1 - c_3)/c_1$ and $2p > (c_3 - c_2)/c_2$, then $C_1 > C_3 \geq C_2$.
(iii) If $2p \leq (c_3 - c_2)/c_2$, then $C_3 \geq C_1 > C_2$.

Table 2.7. Expected costs C_i $(i = 1, 2, 3)$ when $c_n = 2 + n$

p	C_1	C_2	C_3
0.5	16.00	12.50	10.00
0.2	6.25	5.67	5.58
0.1	4.94	4.70	5.14
0.01	4.08	4.02	4.95
0.001	4.01	4.01	5.00
0.0001	4.00	4.00	5.00

Table 2.8. Data length L_0 when $C_2 = C_3$ and $c_n = c_0 + n$

p_1	c_0				
	1	2	3	4	5
10^{-3}	346.4	202.6	143.8	111.5	91.1
10^{-4}	3465.7	2027.3	1438.4	1115.7	911.6
10^{-5}	34657.4	20273.3	14384.1	111571.7	91160.7
10^{-6}	346573.6	202732.5	143841.0	1115717.0	911607.0

In general, the probability p of transmission failure of unit data is not constant and depends on its length L and bit error rate p_1. Suppose that $p \equiv 1 - (1 - p_1)^L$, i.e., $L = \log(1 - p)/\log(1 - p_1)$. Then, the above policy is:

(i)′ If $L \geq \log[(c_2 + c_1 - c_3)/c_1]/[2\log(1 - p_1)]$, then $C_1 > C_2 \geq C_3$.
(ii)′ If $\log[(3c_2 - c_3)/(2c_2)] / \log(1 - p_1) < L < \log[(c_2 + c_1 - c_3)/c_1]/[2\log(1 - p_1)]$, then $C_1 > C_3 \geq C_2$.
(iii)′ If $L \leq \log[(3c_2 - c_3)/(2c_2)]/\log(1 - p_1)$, then $C_3 \geq C_1 > C_2$.

Example 2.7. When $c_n = 2 + n$ $(n = 1, 2, 3)$, Table 2.7 presents the expected costs C_i $(i = 1, 2, 3)$ for p. In this case, when $p = p_0 \equiv 1 - \sqrt{2/3} = 0.1835$, C_2 is equal to C_3. If $p > p_0$, then C_3 is smaller than C_2, and *vice versa*.

In addition, when $p = 1 - (1 - p_1)^L$ and $c_n = c_0 + n$, Table 2.8 presents the data length L_0 at which C_2 is equal to C_3, i.e.,

$$L_0 = \frac{\log[c_0/(c_0 + 1)]}{2\log(1 - p_1)}$$

for p_1 and $c_0 = 1, 2, 3, 4$, and 5. For example, when $p_1 = 10^{-4}$ and $c_0 = 2$, if $L > 2028$, then C_3 is smaller than C_2. It is of interest that $p_1 L_0$ is almost constant for a specified c_0, i.e., $p_1 L_0 \approx (1/2) \log[(c_0 + 1)/c_0]$. ∎

2.5 Other Redundant Models

Using the properties of redundant systems, we apply them to the following three redundant models:

(1) Redundant Bits

The BASIC mode data transmission control procedure is one typical method of data transmission on a public transmission line and is simply called basic procedure [45]. To reduce failures of transmissions, data in basic procedure are often divided into some small blocks, each of which has redundant bits such as heading, control characters, bit check character, and *etc.*

If we send one block to a receiver and he or she finds no error, we call it a block transmission success. If a receiver detects some errors by redundant bits, he or she informs it to us, and we send the same block again. The process is repeated until the success of all block transmissions, *i.e.*, transmission success. This is called ARQ scheme [41].

It is assumed that bit errors occur independently of each other, and its rate (BER) is constant p_1 $(0 < p_1 < 1)$ for any transmission. In addition, we divide unit data with length S into N blocks. Then, the error rate of one block is

$$p = 1 - (1 - p_1)^{S/N}. \tag{2.100}$$

In addition, we attach n redundant bits to each block, so that the length of one block is $S/N + n$. We transmit each block successively to a receiver. If some errors of a block are detected, we retransmit it until block transmission success. Let $M(N)$ be the total expected number of blocks that have been transmitted until data transmission success. Because errors of each block occur with probability p in (2.100), we have a renewal equation:

$$M(N) = \sum_{j=0}^{N} \binom{N}{j} p^j (1-p)^{N-j} [M(j) + N] \qquad (N = 1, 2, \dots),$$

where $M(0) \equiv 0$. Solving for $M(N)$,

$$M(N) = \frac{N}{1-p}. \tag{2.101}$$

Thus, the total average length $L(N)$ of transmission data until transmission success is

$$L(N) = \left(\frac{S}{N} + n\right) M(N) = \frac{S + nN}{(1 - p_1)^{S/N}} \qquad (N = 1, 2, \dots). \tag{2.102}$$

Note that $L(N)$ increases with n. However, when n is small, we might not detect some occurrences of errors and cannot trust the accuracy of data transmission even if it succeeds.

Table 2.9. Optimum number N^* and average data length $L(N^*)$

S	$p_1 = 10^{-4}$				$p_1 = 10^{-5}$			
	$n = 64$		$n = 128$		$n = 64$		$n = 128$	
	N^*	$L(N^*)$	N^*	$L(N^*)$	N^*	$L(N^*)$	N^*	$L(N^*)$
1,024	1	1,205	1	1,276	1	1,099	1	1,164
2,048	3	2,398	2	2,552	1	2,159	1	2,221
4,096	5	4,793	4	5,105	2	4,311	1	4,401
8,192	11	9,584	8	10,210	3	8,616	2	8,801
16,384	21	19,167	15	20,417	7	17,230	5	17,591

Example 2.8. We can easily compute the optimum number N^* that minimizes $L(N)$ in (2.102) for specified S, n, and p_1. Table 2.9 presents the optimum N^* and the resulting length $L(N^*)$ for $p_1 = 10^{-4}$, 10^{-5} and $n = 64$, 128. This indicates that N^* increases with p_1 and S and decreases with n. For example, when $S = 2,048$, $n = 64$, and $p_1 = 10^{-4}$, the optimum number is $N^* = 3$, and the average data length is $L(N^*) = 2,398$, that is 17.1% longer than an original data length S. The rate $L(N^*)/S$ decreases slowly with S. ■

(2) Redundant Networks

Consider a network system with two terminals that consists of N ($N \geq 1$) networks (see Fig. 9.9 in Chap. 9): Customers arrive at the system according to an exponential distribution $(1 - e^{-\lambda t})$, and their usage times also have an identical exponential distribution $(1 - e^{-\mu t})$, *i.e.*, this process forms an M/M/N(∞) queueing one. Then, the probability in the steady-state that there are j customers in the system is [46]

$$p_j = \begin{cases} \dfrac{a^j}{j!} p_0 & (0 \leq j \leq N), \\[2ex] \dfrac{N^N \rho^j}{N!} p_0 & (j \geq N), \end{cases} \tag{2.103}$$

where $a \equiv \lambda/\mu$, $\rho \equiv a/N < 1$, and

$$p_0 = 1 / \left[\sum_{j=0}^{N-1} \frac{a^j}{j!} + \frac{a^N}{(N-1)!(N-a)} \right].$$

We define the probability that customers can use a network without waiting, *i.e.*, the availability of the system is

Table 2.10. Optimum number N^* and system efficiency $C(N^*)/c_1$ when $a = 0.5$

c_0/c_1	N^*	$C(N^*)/c_1$
0.1	1	2.20
0.2	1	2.40
0.5	2	2.78
1.0	2	3.33
2.0	2	4.44
5.0	2	7.78
10.0	3	13.20

$$A(N) \equiv \sum_{j=0}^{N-1} p_j. \tag{2.104}$$

Let $c_1 N + c_0$ be the construction cost for system with N networks. By arguments similar to those in **(1)** of Sect. 2.1.1, we give a system efficiency as

$$C(N) = \frac{c_1 N + c_0}{\sum_{j=0}^{N-1} p_j} \qquad (N = 1, 2, \dots). \tag{2.105}$$

From the inequality $C(N + 1) - C(N) \geq 0$, an optimum network N^* to minimize $C(N)$ is given by a minimum that satisfies

$$\frac{1}{p_N} \sum_{j=0}^{N-1} p_j - N \geq \frac{c_0}{c_1} \qquad (N = 1, 2, \dots). \tag{2.106}$$

It can be easily seen that if p_N decreases strictly with N, then the left-hand side of (2.106) increases strictly with N and tends to ∞ as $N \to \infty$ when $p_N \to 0$ because

$$\frac{1}{p_N} \sum_{j=0}^{N-1} p_j - N > \frac{p_0}{p_N} - 1 \qquad \text{for } N \geq 2.$$

Example 2.9. Table 2.10 presents the optimum N^* and the resulting efficiency $C(N^*)/c_1$ for c_0/c_1 when $a = 0.5$, *i.e.*, $1/\lambda = 2/\mu$, that means that the mean arrival time of customers is two times their mean usage time. In this case, there exits a finite N^* always exists because $p_N \to 0$ as $N \to \infty$. This indicates that the optimum N^* increases slowly as c_0/c_1 increases. ∎

(3) N Copies

One of the most important things in modern societies is the diversification of information and risks. We have to take some copies of important goods to prevent their loss and store them separately in other places. It is assumed that the probabilities of losing all N copies and of at least one of N copies being stolen are p^N and $1 - q^N$, respectively, where $1 \geq q > p > 0$. In addition, we introduce the following costs: A storage cost of N copies is $c_1 N + c_0$, c_2 is the cost of losing all copies and c_3 is the cost of at least one copy being stolen.

We give the total expected cost with N copies as

$$C(N) = c_1 N + c_0 + c_2 p^N + c_3(1 - q^N) \qquad (N = 1, 2, \dots). \qquad (2.107)$$

From the inequality $C(N + 1) - C(N) \geq 0$, an optimum number N^* that minimizes $C(N)$ is a minimum such that

$$\frac{c_1 + c_3 q^N (1 - q)}{p^N (1 - p)} \geq c_2 \qquad (N = 1, 2, \dots). \qquad (2.108)$$

The left-hand side of (2.108) increases strictly with N to ∞. Thus, there exists a finite and unique minimum N^* $(1 \leq N^* < \infty)$ that satisfies (2.108). For example, when $p = 0.2$, $q = 0.99$, $c_2/c_1 = 500$, and $c_3/c_1 = 100$, the optimum number N^* that minimizes $C(N)$ is $N^* = 4$.

3

Partition Policies

There have been many reliability models in our daily life whose performance improves by redundancy described well in Chap. 2. On the other hand, we have met with some interesting models in our studies whose characteristic values improve by partitioning their functions suitably. For example, some performance evaluations of computer systems can improve by partitioning their functions, although increase in costs, times, or overheads may be incurred through partitions. Such problems take their theoretical origin from the basic inspection policy [1, p. 201]. A typical model of partition problems in modern societies is the diversification of information and risks. One of the most important problems from classical reliability theory is the optimum allocation of redundancy subject to some constraints [2, 47].

In Sect. 3.1, we look back to maintenance policies for the periodic inspection model [1, p. 224] and three replacement models with a finite working time S [1, p. 241]. The expected costs for each model when an interval S is partitioned equally into N parts are obtained. It would be more difficult to analyze theoretically optimum policies for discrete problems than those for continuous problems. First, the discrete problem with variable N that minimizes the expected cost converts to the continuous problem with variable $T \equiv S/N$. An optimum number N^* that minimizes the expected cost is derived by using the partition method shown in this section. It is of great interest that the summation of integers from 1 to N plays an important role in obtaining an optimum N^*. Such types of the summation will also appear in (2.85) of Sect. 2.2, Example 5.4 of Sect. 5.5, and (10.57) and (10.67) of Sect. 10.4.

In Sect. 3.2, we introduce four partition models that have exist in our recent studies [48]: (1) Backup policy for a hard disk, (2) job partition, (3) garbage collection, and (4) network partition. Optimum numbers N^* for each model are derived by using the partition method. Finally, in Sect. 3.3, we propose a job partition model as one typical example of computer systems, where a job is partitioned into N tasks with signatures and is executed on two processors. An optimum number N^* that minimizes the mean time to

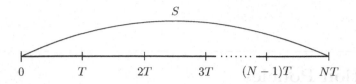

Fig. 3.1. Finite time S with periodic N intervals

complete a job is easily given by the summation of integers from 1 to N in exponential cases.

In other fields of operations research and management science, there would exist similar partition models whose performance improves by partitioning their functions. The techniques and results derived in this chapter would be a good guide to further studies of such models.

3.1 Maintenance Models

Some units would be operating for a finite interval $[0, S]$. The most important maintenance policy for such units is when to check them for the inspection model and when to replace them for the replacement models. Suppose that an interval S is partitioned equally into N ($N = 1, 2, \ldots$) parts. For the inspection policy, a unit is checked, and for the replacement policies, it is replaced periodically at planned times kT ($k = 1, 2, \ldots, N$), where $NT = S$ (Fig. 3.1). The expected costs for each model are obtained, and optimum numbers N^* that minimize them are derived by using the partition method. Optimum maintenance policies for usual replacement models, preventive maintenance models, and inspection models were summarized [49]. Sequential maintenance policies for a finite time span will be discussed in Chap. 4.

3.1.1 Inspection Polices

Suppose that a unit has to be operating for a finite interval $[0, S]$ ($0 < S < \infty$) and fails according to a failure distribution $F(t)$ with a finite mean $\mu \equiv \int_0^\infty \overline{F}(t) \mathrm{d}t < \infty$, where $\overline{F}(t) \equiv 1 - F(t)$. Then, an interval S is partitioned equally into N parts. The unit is checked at periodic times kT ($k = 1, 2, \ldots, N$) as a periodic inspection policy, where $NT = S$. It is assumed that any failures of the unit are always detected only through such checks.

Let c_I be the cost for each check and c_D be the cost per unit of time elapsed between a failure and its detection at the next check. Then, the total expected cost until failure detection or time S is, from [1, p. 203],

$$C(N) = \sum_{k=0}^{N-1} \int_{kT}^{(k+1)T} \{c_I(k+1) + c_D[(k+1)T - t]\}\,dF(t) + c_I N\overline{F}(NT)$$

$$= \left(c_I + \frac{c_D S}{N}\right) \sum_{k=0}^{N-1} \overline{F}\left(\frac{kS}{N}\right) - c_D \int_0^S \overline{F}(t)\,dt \quad (N = 1, 2, \ldots). \quad (3.1)$$

We find an optimum number N^* that minimizes $C(N)$. It is clearly seen that $\lim_{N \to \infty} C(N) = \infty$ and

$$C(1) = c_I + c_D \int_0^S F(t)\,dt. \quad (3.2)$$

Thus, there exists a finite number N^* $(1 \le N^* < \infty)$ that minimizes $C(N)$.

In particular, assume that the failure time is exponential, i.e., $F(t) = 1 - e^{-\lambda t}$. In this case, the expected cost $C(N)$ in (3.1) is rewritten as

$$C(N) = \left(c_I + \frac{c_D S}{N}\right) \frac{1 - e^{-\lambda S}}{1 - e^{-\lambda S/N}} - \frac{c_D}{\lambda}(1 - e^{-\lambda S}) \quad (N = 1, 2, \ldots). \quad (3.3)$$

Forming the inequality $C(N+1) - C(N) \ge 0$,

$$\frac{e^{-\lambda S/(N+1)} - e^{-\lambda S/N}}{[1 - e^{-\lambda S/(N+1)}]/N - [1 - e^{-\lambda S/N}]/(N+1)} \ge \frac{c_D S}{c_I} \quad (N = 1, 2, \ldots). \quad (3.4)$$

Using two approximations by a Taylor expansion that

$$e^{-\lambda S/(N+1)} - e^{-\lambda S/N} \approx \frac{\lambda S}{N} - \frac{\lambda S}{N+1},$$

$$\frac{1 - e^{-\lambda S/(N+1)}}{N} - \frac{1 - e^{-\lambda S/N}}{N+1} \approx \frac{1}{2(N+1)}\left(\frac{\lambda S}{N}\right)^2 - \frac{1}{2N}\left(\frac{\lambda S}{N+1}\right)^2,$$

(3.4) becomes simply

$$\sum_{j=1}^{N} j = \frac{N(N+1)}{2} \ge \frac{\lambda S}{4}\frac{c_D S}{c_I}. \quad (3.5)$$

Thus, an approximate number $\widetilde{N}$ that minimizes $C(N)$ in (3.3) is easily given by (3.5).

Note that the summation of integers from 1 to N, i.e., $N(N+1)/2 = 1, 3, 6, 10, 15, 21, 28, 36, 45, 55, \ldots$, plays an important role in obtaining optimum partition policies for some models. Such a type of equations appears in the optimization problem that minimizes objective functions

$$C(N) = AN + \frac{B}{N} \quad (N = 1, 2, \ldots),$$

and

$$C(N) = A^N B^{1/N} \qquad (N = 1, 2, \dots),$$

where parameters A and B are constant. In general, an optimum N^* that minimizes

$$C(N) = AN^\beta + \frac{B}{N^\beta}$$

for positive $\beta > 0$ is given by a unique minimum such that

$$\left(2 \sum_{j=1}^{N} j\right)^\beta \geq \frac{B}{A}.$$

Furthermore, setting $T = S/N$ in (3.3), it follows that

$$C(T) = (c_I + c_D T) \frac{1 - e^{-\lambda S}}{1 - e^{-\lambda T}} - \frac{c_D}{\lambda}(1 - e^{-\lambda S}). \qquad (3.6)$$

Differentiating $C(T)$ with respect to T and setting it equal to zero,

$$e^{\lambda T} - (1 + \lambda T) = \frac{\lambda c_I}{c_D}, \qquad (3.7)$$

whose left-hand side increases strictly from 0 to ∞. Thus, there exists a finite and unique $\widetilde{T}$ $(0 < \widetilde{T} < \infty)$ that satisfies (3.7). Note that $\widetilde{T}$ gives the optimum checking time of the periodic inspection policy for an infinite time span in an exponential case [1, p. 204].

Therefore, we have the following optimum partition method:

(1) If $\widetilde{T} < S$ and $[S/\widetilde{T}] \equiv N$, then calculate $C(N)$ and $C(N+1)$ from (3.3), where $[x]$ denotes the greatest integer contained in x. If $C(N) \leq C(N+1)$, then $N^* = N$, and conversely, if $C(N) > C(N+1)$, then $N^* = N+1$.
(2) If $\widetilde{T} \geq S$, then $N^* = 1$.

Example 3.1. Table 3.1 presents the approximate checking time $\widetilde{T}$, the optimum checking number N^* and time $T^* = S/N^*$, the resulting expected cost $C(N^*)/c_D$ in (3.3), and the approximate number $\widetilde{N}$ that satisfies (3.5) for $S = 100$, 200, and $c_I/c_D = 2$, 5, 10, and 25 when $\lambda = 0.01$. It is of interest that the approximate number is $\widetilde{N} \leq N^*$, however, it is almost the same as the optimum. As a result, it would be sufficient in actual fields to adopt the checking number $\widetilde{N}$ for a finite time span when the failure time is exponential. ∎

3.1.2 Replacement Policies

A unit has to be operating for a finite interval $[0, S]$, *i.e.*, the working time of a unit is given by a specified value S. To maintain a unit, an interval S is partitioned equally into N parts in which it is replaced at periodic times kT $(k = 1, 2, \dots, N)$, where $NT \equiv S$. Then, we consider the replacement with minimal repair at failure, the block replacement, and the simple replacement.

Table 3.1. Approximate time $\widetilde{T}$, optimum number N^* and time T^*, expected cost $C(N^*)/c_D$, and approximate number $\widetilde{N}$ when $\lambda = 0.01$

S	c_I/c_D	$\widetilde{T}$	N^*	T^*	$C(N^*)/c_D$	$\widetilde{N}$
	2	19.355	5	20.000	13.506	5
	5	30.040	3	33.333	22.269	3
100	10	41.622	2	50.000	33.180	2
	25	63.271	2	50.000	57.278	1
	2	19.355	10	20.000	18.475	10
	5	30.040	7	28.571	30.336	6
200	10	41.622	5	40.000	44.671	4
	25	63.271	3	66.667	76.427	3

(1) Periodic Replacement with Minimal Repair

The unit is replaced at periodic times kT $(k = 1, 2, \ldots, N)$, and any units are as good as new at each replacement. When the unit fails between replacements, only minimal repair is made, and hence, its failure rate remains undisturbed by any repair of failures [1, p. 96]. It is assumed that the repair and replacement times are negligible. Suppose that the failure time of the unit has a density function $f(t)$ and a failure distribution $F(t)$, i.e., $f(t) \equiv dF(t)/dt$. Then, the failure rate is $h(t) \equiv f(t)/\overline{F}(t)$, and its cumulative hazard rate is $H(t) \equiv \int_0^t h(u)du$, i.e., $\overline{F}(t) \equiv 1 - F(t) = e^{-H(t)}$, that represents the expected number of failures during $(0, t]$.

Let c_M be the cost for minimal repair and c_T be the cost for planned replacement at time kT. Then, the expected cost for one interval $[(k-1)T, kT]$ is [2]

$$\widetilde{C}_1(1) \equiv c_M H(T) + c_T = c_M H\left(\frac{S}{N}\right) + c_T. \tag{3.8}$$

Thus, the total expected cost for a finite interval $[0, S]$ is

$$C_1(N) \equiv N\widetilde{C}_1(1) = N\left[c_M H\left(\frac{S}{N}\right) + c_T\right] \quad (N = 1, 2, \ldots). \tag{3.9}$$

Clearly, $\lim_{N \to \infty} C_1(N) = \infty$ and

$$C_1(1) = c_M H(S) + c_T. \tag{3.10}$$

Thus, there exists a finite number N^* $(1 \le N^* < \infty)$ that minimizes $C_1(N)$. Forming the inequality $C_1(N+1) - C_1(N) \ge 0$,

$$\frac{1}{NH\left(\frac{S}{N}\right) - (N+1)H\left(\frac{S}{N+1}\right)} \ge \frac{c_M}{c_T} \quad (N = 1, 2, \ldots). \tag{3.11}$$

When the failure time has a Weibull distribution, i.e., $H(t) = \lambda t^m$ $(m > 1)$, (3.11) becomes

$$\frac{1}{[1/N^{m-1}] - [1/(N+1)^{m-1}]} \geq \frac{c_M \lambda S^m}{c_T} \qquad (N = 1, 2, \ldots). \qquad (3.12)$$

The left-hand side of (3.12) increases strictly with N to ∞ because $[1/x]^\alpha - [1/(x+1)]^\alpha$ decreases strictly with x for $1 \leq x < \infty$ and $\alpha > 0$. Thus, there exists a finite and unique minimum N^* $(1 \leq N^* < \infty)$ that satisfies (3.12). In particular, when $m = 2$, i.e., $H(t) = \lambda t^2$, (3.12) is

$$\frac{N(N+1)}{2} \geq \frac{c_M \lambda S^2}{2c_T}, \qquad (3.13)$$

that agrees with the type of inequality (3.5).

To obtain an optimum N^*, setting $T \equiv S/N$ in (3.9), it follows that

$$C_1(T) = S \left[\frac{c_M H(T) + c_T}{T} \right]. \qquad (3.14)$$

Thus, the problem of minimizing $C_1(T)$ corresponds to that of the standard replacement with minimal repair for an infinite time span. Many discussions on such optimum policies have been held [1, p. 101]: Differentiating $C_1(T)$ with respect to T and setting it equal to zero,

$$Th(T) - H(T) = \frac{c_T}{c_M}. \qquad (3.15)$$

When the failure rate $h(t)$ increases strictly, the left-hand side of (3.15) also increases strictly, and hence, a solution $\widetilde{T}$ to satisfy (3.15) is unique if it exists. Therefore, using the partition method in Sect. 3.1.1, we can get the optimum number N^*.

(2) Block Replacement

Suppose that a unit is always replaced at any failures between replacements. This is called block replacement and has been studied [1, p. 117]: Let $M(t)$ be the renewal function of distribution $F(t)$, i.e., $M(t) \equiv \sum_{j=1}^{\infty} F^{(j)}(t)$, where $F^{(j)}(t)$ is the jth Stieltjes convolution of $F(t)$, and $F^{(j)}(t) \equiv \int_0^t F^{(j-1)}(t - u) dF(u)$ $(j = 1, 2, \ldots)$ and $F^{(0)}(t) \equiv 1$ for $t \geq 0$, 0 for $t < 0$, that is, $M(t)$ represents the expected number of failed units during $(0, t]$.

Let c_F be the replacement cost for a failed unit and c_T be the cost for planned replacement at time kT. Then, the expected cost for one interval $[(k-1)T, kT]$ is [2]

$$\widetilde{C}_2(1) \equiv c_F M(T) + c_T = c_F M \left(\frac{S}{N} \right) + c_T.$$

Thus, the total expected cost for a finite interval $[0, S]$ is

$$C_2(N) \equiv N\tilde{C}_2(1) = N\left[c_F M\left(\frac{S}{N}\right) + c_T\right] \qquad (N = 1, 2, \dots). \tag{3.16}$$

Forming the inequality $C_2(N + 1) - C_2(N) \geq 0$,

$$\frac{1}{NM(\frac{S}{N}) - (N+1)M\left(\frac{S}{N+1}\right)} \geq \frac{c_F}{c_T} \qquad (N = 1, 2, \dots). \tag{3.17}$$

When the failure time has a gamma distribution of order 2, i.e., $\overline{F}(t) = (1 + \lambda t)e^{-\lambda t}$,

$$M(t) = \frac{\lambda t}{2} - \frac{1}{4}(1 - e^{-2\lambda t}).$$

Thus, (3.17) is

$$\frac{4}{(N+1)[1 - e^{-2\lambda S/(N+1)}] - N[1 - e^{-2\lambda SN}]} \geq \frac{c_F}{c_T}. \tag{3.18}$$

Using the approximation $e^{-a} \approx 1 - a + a^2/2$, (3.18) is simply rewritten as

$$\frac{N(N+1)}{2} \geq \left(\frac{\lambda S}{2}\right)^2 \frac{c_F}{c_T}, \tag{3.19}$$

that agrees with the type of inequality (3.5).

Furthermore, setting $T \equiv S/N$ in (3.16),

$$C_2(T) = S\left[\frac{c_F M(T) + c_T}{T}\right], \tag{3.20}$$

that corresponds to the standard block replacement. Let $m(t)$ be the renewal density function of $F(t)$, i.e., $m(t) \equiv dM(t)/dt$. Then, differentiating $C_2(T)$ with respect to T and setting it equal to zero,

$$Tm(T) - M(T) = \frac{c_T}{c_F}. \tag{3.21}$$

Therefore, using the partition method, we can get the optimum policy.

(3) Simple Replacement

Suppose that a unit is always replaced at times kT ($k = 1, 2, \dots, N$), but it is not replaced instantly at failure, and hence, it remains in a failed state for the time interval from a failure to its detection [1, p. 120].

Let c_D be the cost for the time elapsed between a failure and its detection per unit of time and c_T be the cost for planned replacement at time kT. Then, the expected cost for one interval $[(k-1)T, kT]$ is

$$\widetilde{C}_3(1) = c_D \int_0^T F(t)\,dt + c_T = c_D \int_0^{S/N} F(t)\,dt + c_T. \qquad (3.22)$$

Thus, the total expected cost for a finite interval $[0, S]$ is

$$C_3(N) \equiv N\widetilde{C}_3(1) = N\left[c_D \int_0^{S/N} F(t)\,dt + c_T\right] \qquad (N = 1, 2, \dots). \quad (3.23)$$

Forming the inequality $C_3(N+1) - C_3(N) \geq 0$,

$$\frac{1}{N \int_0^{S/N} F(t)dt - (N+1) \int_0^{S/(N+1)} F(t)dt} \geq \frac{c_D}{c_T} \quad (N = 1, 2, \dots). \quad (3.24)$$

In particular, when $F(t) = 1 - e^{-\lambda t}$, (3.24) is

$$\frac{1}{(N+1)(1 - e^{-\lambda S/(N+1)}) - N(1 - e^{-\lambda S/N})} \geq \frac{c_D}{\lambda c_T} \quad (N = 1, 2, \dots). \quad (3.25)$$

Using the same approximation as in **(2)**, (3.25) is

$$\frac{N(N+1)}{2} \geq \left(\frac{\lambda S}{2}\right)^2 \frac{c_D}{\lambda c_T}. \qquad (3.26)$$

Furthermore, setting $T \equiv S/N$ in (3.23),

$$C_3(T) = S\left[\frac{c_D \int_0^T F(t)dt + c_T}{T}\right]. \qquad (3.27)$$

Differentiating $C_3(T)$ with respect to T and setting it equal to zero,

$$TF(T) - \int_0^T F(t)\,dt = \frac{c_T}{c_D}. \qquad (3.28)$$

Noting that the left-hand side of (3.28) increases strictly from 0 to μ, there exists a finite and unique $\widetilde{T}$ that satisfies (3.28), if $\mu > c_T/c_D$. Therefore, using the partition method, we can get the optimum policy.

In general, the above results for the three replacements are summarized as follows: The total expected cost for a finite interval $[0, S]$ is

$$C(N) = N\left[c_i \Phi\left(\frac{S}{N}\right) + c_T\right] \qquad (N = 1, 2, \dots), \qquad (3.29)$$

where $\Phi(t)$ is $H(t)$, $M(t)$, and $\int_0^t F(u)du$ and c_i is c_M, c_F, c_D for the respective replacements. Forming the inequality $C(N+1) - C(N) \geq 0$,

$$\frac{1}{N\Phi(\frac{S}{N}) - (N+1)\Phi(\frac{S}{N+1})} \geq \frac{c_i}{c_T} \qquad (N = 1, 2, \dots). \qquad (3.30)$$

Setting $T = S/N$ in (3.29),

$$C(T) = S \left[\frac{c_i \Phi(T) + c_T}{T} \right], \qquad (3.31)$$

and differentiating $C(T)$ with respected to T and setting it equal to zero,

$$T\Phi'(T) - \Phi(T) = \frac{c_T}{c_i} \quad \text{or} \quad \int_0^T t \, d\Phi'(t) = \frac{c_T}{c_i}. \qquad (3.32)$$

If there exists a solution to $\tilde{T}$ to (3.32), then we can get an optimum number N^* for each replacement by using the partition method.

In this section, we make no mention of age replacement, where a unit is replaced at time T or at failure, whichever occurs first [1, p. 69]. We can obtain an optimum age replacement policy for a finite time span by a similar method as follows: We partition a whole working time S into N equal parts, i.e., $NT \equiv S$, and derive an optimum replacement time for one interval $(0, T]$. Using the partition method, we can determine an optimum number N^*. If a unit is replaced at time T_0 ($0 < T_0 \leq T$), then we may reconsider the same replacement policy for the remaining interval $(0, S - T_0]$.

3.2 Partition Models

There exist many reliability models whose performance is improved by partitioning their functions. This section presents four partition models in computer and information systems and analyzes them by applying the partition method to them. The recovery problems when checkpoints should be placed for a finite execution time S will be similarly discussed in Sect. 7.1.

(1) Backup Policy for A Hard Disk

A hard disk stores a variety of files that are frequently updated by adding or deleting them. These files might occasionally be lost due to human errors or failures of some hardware devices. The backup policy for the hard disk with a whole volume S was studied [50, 51]: The backup operation is executed when the files of S/N ($N = 1, 2, \dots$) are created or updated. It is assumed that the time needed for the backup operation is proportional to S/N, i.e., its time is aS/N ($0 < a < 1$). Furthermore, the time considered here is measured by the time until a whole volume S is executed or is used fully, i.e., the measure of S may be expressed as the time. The problem of scheduling the checkpoints to save the completed work of a given job was discussed [52].

Suppose that the hard disk fails according to an exponential distribution $(1 - e^{-\lambda t})$. In this case, the total files or data in the hard disk after the previous backup are lost due to its failure. Let c_B be the cost for the backup operation and c_D be the cost per unit of time for losing files. Then, the expected cost $C(1)$ for one backup interval is given by a renewal function:

$$C(1) = c_B e^{-\lambda(a+1)S/N} + \int_0^{S/N} [c_D x + C(1)]\lambda e^{-\lambda x}\, dx$$

$$+ \int_{S/N}^{(a+1)S/N} \left[c_B + \frac{c_D S}{N} + C(1) \right] \lambda e^{-\lambda x}\, dx. \tag{3.33}$$

Thus, the total expected cost until a whole volume S is successfully executed is

$$C(N) \equiv NC(1) = N c_B e^{\lambda a S/N} + \frac{N c_D}{\lambda} e^{\lambda a S/N}(e^{\lambda S/N} - 1) - c_D S$$

$$(N = 1, 2, \dots). \tag{3.34}$$

Using the approximation $e^a \approx 1 + a$, the expected cost is simply

$$\widetilde{C}(N) = c_B(N + \lambda a S) + c_D \frac{\lambda a S^2}{N}. \tag{3.35}$$

Thus, an optimum number $\widetilde{N}$ that minimizes (3.35) is given by a unique minimum of the inequality

$$\frac{N(N+1)}{2} \geq \frac{c_D \lambda a S^2}{2 c_B}, \tag{3.36}$$

that agrees with the type of inequality (3.5).

Setting $T \equiv S/N$, (3.34) becomes

$$C(T) \equiv S \left[\frac{c_B e^{\lambda a T} + c_D e^{\lambda a T}(e^{\lambda T} - 1)/\lambda}{T} - c_D \right]. \tag{3.37}$$

Differentiating $C(T)$ with T and setting it equal to zero,

$$\lambda T(c_B \lambda a + c_D e^{\lambda T}) + c_D(\lambda a T - 1)(e^{\lambda T} - 1) = \lambda c_B, \tag{3.38}$$

whose left-hand side increases strictly from 0 to ∞. Thus, there exists a finite and unique $\widetilde{T}$ that satisfies (3.38).

Therefore, using the partition method, we can get an optimum number N^* that minimizes the expected cost $C(N)$ in (3.34). It can be easily seen that an optimum N^* increases with a because $\widetilde{T}$ decreases with a. Generally, the shorter the backup time would be, the more frequently the backup should be done. Moreover, when S is sufficiently large, we may do the backup every time at $\widetilde{T}$, irrespective of N and S.

(2) Job Partition

A job is executed on a microprocessor (μP) and is partitioned into small tasks. If a job is not partitioned, it has to be executed again from the beginning when its process has failed. Suppose that S is an original process time of a job. Then, we partition a job equally into N tasks ($N = 1, 2, \ldots$). Each task has the process time S/N and is executed on a μP. It is assumed that a μP fails according to an exponential distribution ($1 - e^{-\lambda S/N}$), and each task is executed again from the beginning. The process of a job succeeds when all processes of N tasks are completed. Optimum problems of determining retrial numbers for such stochastic models will be discussed in Chap. 6.

Let c_P be the prepared time for the execution of one task and c_F be the prepared time for reexecution of one task when its process has failed. Then, the mean process time until one task is completed is given by a renewal equation:

$$L(1) = \left(\frac{S}{N} + c_P\right) e^{-\lambda S/N} + \left(\frac{S}{N} + c_P + c_F + L(1)\right)(1 - e^{-\lambda S/N}), \quad (3.39)$$

and solving it for $L(1)$,

$$L(1) = \left(\frac{S}{N} + c_P\right) e^{\lambda S/N} + c_F(e^{\lambda S/N} - 1). \quad (3.40)$$

Thus, the mean time to the completion of the job is

$$L(N) \equiv NL(1) = (S + Nc_P)e^{\lambda S/N} + Nc_F(e^{\lambda S/N} - 1) \quad (N = 1, 2, \ldots). \quad (3.41)$$

We find an optimum number N^* that minimizes $L(N)$. Using the approximation of $e^a \approx 1 + a$, (3.41) is

$$\widetilde{L}(N) = S + Nc_P + \frac{\lambda S^2}{N} + (c_P + c_F)\lambda S. \quad (3.42)$$

Thus, an optimum $\widetilde{N}$ to minimize $\widetilde{L}(N)$ is given by a unique minimum that satisfies

$$\frac{N(N+1)}{2} \geq \frac{\lambda S^2}{2c_P}, \quad (3.43)$$

that agrees with the type of inequality (3.5).

Furthermore, setting $T \equiv S/N$, (3.41) is

$$L(T) = S\left[\left(1 + \frac{c_P}{T}\right) e^{\lambda T} + \frac{c_F}{T}(e^{\lambda T} - 1)\right]. \quad (3.44)$$

Differentiating $L(T)$ with respect to T and setting it equal to zero,

$$\lambda T^2 + (c_P + c_F)\lambda T - c_F(1 - e^{-\lambda T}) = c_P, \quad (3.45)$$

whose left-hand side increases strictly from 0 to ∞. Thus, there exists a finite and unique $\widetilde{T}$ that satisfies (3.45). Therefore, applying the partition method to this model, we can get an optimum N^* that minimizes the mean time $L(N)$ in (3.41).

(3) Garbage Collection

A database of a computer system has to be operating for a finite interval $(0, S]$. However, after some operations, storage areas are not in good order due to additions or deletions of data. To use a storage area effectively and to improve processing efficiently, garbage collections are done at periodic times kT $(k = 1, 2, \ldots, N)$, where $NT = S$ [53,54]. Some optimum garbage collection policies that are done at a planned time and an update number were summarized [3, p. 131], using the results of shock and damage models.

Suppose that an amount of garbage arises according to an identical exponential distribution $(1 - e^{-\lambda x/T})$ with a mean T/λ for each interval $((k-1)T, kT]$ $(k = 1, 2, \ldots, N)$. Let $c_1 + c_2(x)$ be the cost function required for the garbage collection when the amount of garbage is x at time kT, where $c_2(0) \equiv 0$. Then, the expected cost for one interval is

$$C(1) = \int_0^\infty [c_1 + c_2(x)] \, \mathrm{d}(1 - e^{-\lambda x/T}).$$

Thus, the total expected cost during $(0, S]$ is

$$C(N) \equiv NC(1) = N \left[c_1 + \int_0^\infty c_2(x) \, \mathrm{d}(1 - e^{-\lambda x/T}) \right] \quad (N = 1, 2, \ldots).$$
(3.46)

In particular, when $c_2(x) = c_2 x^2 + c_3 x$,

$$C(N) = c_1 N + \frac{2c_2 S^2}{N\lambda^2} + \frac{c_3 S}{\lambda} \quad (N = 1, 2, \ldots).$$
(3.47)

Forming the inequality $C(N+1) - C(N) \geq 0$,

$$\frac{N(N+1)}{2} \geq \frac{c_2}{c_1} \left(\frac{S}{\lambda} \right)^2.$$
(3.48)

Thus, an optimum N^* that minimizes $C(N)$ in (3.47) is given by a unique minimum that satisfies (3.48).

(4) Network Partition

Consider a network with two terminals. Four algorithms for computing the network reliability were compared [55], and a network reliability algorithm accounting for imperfect nodes was proposed [56].

Suppose that a network with two terminals and length S is partitioned equally and independently into N links and $(N-1)$ nodes $(N = 1, 2, \ldots)$ (Fig. 3.2). It is assumed that the respective reliabilities of each link and node are $e^{-\lambda S/N}$ and q $(0 < q < 1)$ independently. Then, the mean length in which the links of a network are normal is

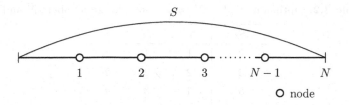

Fig. 3.2. Network with $N-1$ nodes

$$\sum_{j=1}^{N} \frac{jS}{N} \binom{N}{j} (e^{-\lambda S/N})^j (1 - e^{-\lambda S/N})^{N-j} = Se^{-\lambda S/N}.$$

Thus, the mean length of a network with imperfect nodes is

$$L_1(N) = q^{N-1} Se^{-\lambda S/N} \qquad (N = 1, 2, \dots). \qquad (3.49)$$

Therefore, an optimum partition number N^* that maximizes $L_1(N)$ is easily given by a unique minimum that satisfies

$$\frac{N(N+1)}{2} \geq \frac{\lambda S}{-2 \log q}. \qquad (3.50)$$

Next, when a network is partitioned into N links, we denote its complexity as $\log_2 N$ and its reliability as $\exp(-\alpha \log_2 N)$ for $\alpha > 0$ that will be defined in Chap. 9. Then, replacing q^{N-1} in (3.49) with $\exp(-\alpha \log_2 N)$ formally, the mean length of a network with complexity is

$$L_2(N) = S \exp(-\alpha \log_2 N - \lambda S/N) \qquad (N = 1, 2, \dots). \qquad (3.51)$$

An optimum N^* to maximize $L_2(N)$ is a unique minimum that satisfies

$$N(N+1) \log_2 \frac{N+1}{N} \geq \frac{\lambda S}{\alpha} \qquad (N = 1, 2, \dots). \qquad (3.52)$$

The left-hand side of (3.52) increases strictly with N to ∞ and is within the upper bound such that

$$N+1 \leq (N+1) \log_2 \left(1 + \frac{1}{N}\right)^N < (N+1) \log_2 e. \qquad (3.53)$$

Thus, we can easily get a lower bound $\underline{N}$ that satisfies

$$N+1 > \frac{\lambda S}{\alpha \log_2 e}, \qquad (3.54)$$

and an upper bound $\overline{N} \geq (\lambda S)/\alpha - 1$.

Table 3.2. Optimum number N^* and approximate numbers $\overline{N}$ and $\underline{N}$

$\lambda S/\alpha$	N^*	$\underline{N}$	$\overline{N}$
2	1	1	1
3	2	2	2
5	4	3	4
10	7	6	9
20	14	13	19
50	35	34	49
100	69	69	99

Example 3.2.   Table 3.2 presents the optimum number N^*, lower bound $\underline{N}$ and upper bound $\overline{N}$ for $\lambda S/\alpha$. This indicates that $\underline{N}$ gives a better lower bound of N^*, i.e., $N^* \approx (\lambda S/\alpha) \times 0.7$. An upper bound $\overline{N} = (\lambda S)/\alpha - 1$ is easily computed and would be useful for small $\lambda S/\alpha$. ■

3.3 Job Execution with Signature

We can extend the job partition model in (2) to a more generalized model [57]: Microprocessors (μPs) have been widely used in many practical fields, and the demand for improvement of their reliabilities has increased. μPs often fail through some errors due to noises and change in the environment. It is imperative to detect their errors by all means because they require high reliability and safety. Checkpoints that compare and store states or signatures as recovery techniques are well-known schemes for error detection [58, 59], and their optimum policies will be summarized systematically in Chap. 7. A signature is the characteristic information that can be collected by computing the bus information in the operating state [60–62]. A parity code and checkpointing data are also kinds of such signatures. Recently, watchdog processors that detect errors by comparing signatures and computing results have been widely used [63, 64].

Suppose that a μP consists of two processors that execute the same job. The job is partitioned into N tasks with signatures. If the job is not partitioned, it has to be executed again from the beginning when errors have occurred. Consequently, this may increase the total time of job execution. If the job is partitioned into some tasks, then two processors execute the same task with signatures that are compared when all processes are completed. If signatures do not match each other, two processors execute again from the beginning. If they match, then the two processors proceed to the next task.

We are interested in knowing under what conditions the total time of job execution would be reduced by partitioning the job into tasks. For this purpose, we obtain the mean time to complete the job successfully, using the techniques of Markov renewal processes [1, p. 28], and find an optimum number of tasks that minimizes it.

We consider the following job execution with signatures:

(1) A μP consists of two processors that execute the same job.
(2) A job is partitioned into N ($N = 1, 2, \dots$) tasks with signatures that are executed sequentially. The processing time of each task has an identical distribution $A(t)$ with a mean a/N. Signatures are compared with each other when each task ends. The comparison time has a general distribution $B_1(t)$ with a mean b_1.
   (a) When the signatures do not match, the process is not correct. In this case, the task is executed again after a delay time that has a general distribution $D_1(t)$ with a mean d_1.
   (b) When the signatures match, the next task is executed. After the processes of N tasks are completed, all results of tasks are compared because only the signatures are compared and the process results may not be correct. The comparison time has a general distribution $B_2(t)$ with a mean b_2. When its comparison matches, the process result of the job is correct. On the other hand, when its comparison does not match, the process result is not correct. In this case, the job executes again from the beginning after a delay time that has a general distribution $D_2(t)$ with a mean d_2. The probability that its comparison matches is q ($0 < q \le 1$).
(3) Errors of one processor in the execution of tasks occur independently according to an exponential distribution $(1 - e^{-\lambda t})$.
   (a) Some errors are detected by the signatures when the process of each task ends. Undetected errors are detected finally by comparing all results of tasks.
   (b) When errors have occurred, the signatures do not match.
(4) When all processes of N tasks are completed, the job is completed successfully.

Under the above assumptions, we define the following states:

State 0: Process of the job starts.

State j: Process of the jth task is completed ($j = 1, 2, \dots, N$).

State E: Process of the job is completed successfully.

The states defined above form a Markov renewal process, where E is an absorbing state [1, p. 28] (Fig. 3.3).

Let $Q_{ij}(t)$ ($i = 0, 1, \dots, N$; $j = 0, 1, \dots, N, E$) be the mass functions from State i to State j by the probability that after entering State i, the process makes a transition into State j in an amount of time less than or equal to t. Then, we have the following equation:

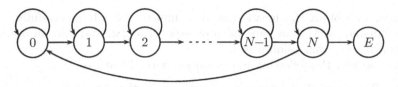

Fig. 3.3. State transition diagram of a Markov renewal process

$$Q_{jj}(t) = \int_0^t (1 - e^{-2\lambda u})\, dA(u) * B_1(t) * D_1(t),$$

$$Q_{jj+1}(t) = \int_0^t e^{-2\lambda u}\, dA(u) * B_1(t) \quad (j = 0, 1, \ldots, N-1),$$

$$Q_{N0}(t) = (1-q)B_2(t) * D_2(t),$$

$$Q_{NE}(t) = qB_2(t), \tag{3.55}$$

where the asterisk denotes the Stieltjes convolution, *i.e.*, $a(t) * b(t) \equiv \int_0^t a(t - u)\,db(u)$.

First, we derive the mean time $\ell_{0E}(N)$ until the job is completed successfully. Let $H_{0E}(t)$ be the first-passage time distribution from State 0 to State E. Then, we have a renewal equation:

$$H_{0E}(t) = \left[\sum_{i=1}^{\infty} Q_{00}^{(i-1)}(t) * Q_{01}(t)\right] * \left[\sum_{i=1}^{\infty} Q_{11}^{(i-1)}(t) * Q_{12}(t)\right]$$

$$* \cdots * \left[\sum_{i=1}^{\infty} Q_{N-1N-1}^{(i-1)}(t) * Q_{N-1N}(t)\right] * [Q_{NE}(t) + Q_{N0}(t) * H_{0E}(t)], \tag{3.56}$$

where $\Phi^{(i)}(t)$ denotes the i-fold Stieltjes convolution of any distribution $\Phi(t)$ with itself, *i.e.*, $\Phi^{(i)}(t) \equiv \Phi^{(i-1)}(t) * \Phi(t) = \int_0^t \Phi^{(i-1)}(t-u)\,d\Phi(u)$ and $\Phi^{(0)}(t) \equiv 1$ for $t \geq 0$.

Let $\Phi^*(s)$ be the Laplace-Stieltjes (LS) transform of any function $\Phi(t)$, *i.e.*, $\Phi^*(s) \equiv \int_0^\infty e^{-st}d\Phi(t)$ for $s > 0$. Then, forming the LS transforms of (3.55) and (3.56), and solving (3.56) for $H_{0E}^*(s)$,

$$H_{0E}^*(s) = \frac{\{Q_{jj+1}^*(s)/[1 - Q_{jj}^*(s)]\}^N q B_2^*(s)}{1 - \{Q_{jj+1}^*(s)/[1 - Q_{jj}^*(s)]\}^N (1-q)B_2^*(s)D_2^*(s)}, \tag{3.57}$$

where

$$Q_{jj}^*(s) = [A^*(s) - A^*(s + 2\lambda)]B_1^*(s)D_1^*(s),$$

$$Q_{jj+1}^*(s) = A^*(s + 2\lambda)B_1^*(s).$$

Thus, the mean time $\ell_{0E}(N)$ is given by

$$\ell_{0E}(N) \equiv \lim_{s \to 0} \frac{1 - H_{0E}^*(s)}{s}$$

$$= \frac{1}{q} \left[\frac{N\{b_1 + d_1[1 - A^*(2\lambda)]\} + a}{A^*(2\lambda)} + b_2 + (1 - q)d_2 \right]$$

$$(N = 1, 2, \dots). \qquad (3.58)$$

Furthermore, we derive the expected number $M(N)$ of task executions until the job is completed successfully. Because the expected number of the ith execution is

$$\sum_{i=1}^{\infty} i[Q_{jj}^*(0)]^{i-1} Q_{jj+1}^*(0) = \frac{1}{1 - Q_{jj}^*(0)} = \frac{1}{A^*(2\lambda)},$$

the total expected number of task executions is

$$M(N) = \frac{N}{A^*(2\lambda)} \sum_{j=1}^{\infty} j(1-q)^{j-1} q = \frac{N}{qA^*(2\lambda)} \qquad (N = 1, 2, \dots). \qquad (3.59)$$

In particular, assume that the process time of each task has an exponential distribution $A(t) = (1 - e^{-Nt/a})$. Then, the mean time $\ell_{0E}(N)$ is, from (3.58),

$$\ell_{0E}(N) = \frac{1}{q} \left[(N + 2\lambda a) \left(\frac{a}{N} + b_1 \right) + 2\lambda d_1 a + b_2 + (1 - q)d_2 \right]$$

$$(N = 1, 2, \dots). \qquad (3.60)$$

Therefore, an optimum number N^* that minimizes $\ell_{0E}(N)$ is easily given by

$$\frac{N(N+1)}{2} \geq \frac{\lambda a^2}{b_1}, \qquad (3.61)$$

that agrees with the type of inequality (3.5). In this case, the total number of task executions is

$$M(N^*) = \frac{1}{q}(N^* + 2\lambda a). \qquad (3.62)$$

Example 3.3. Suppose that the comparison time b_1 of signatures is a unit time, and the mean process time when the job is not partitioned is $a/b_1 = 100$–400, where the parameter a represents the job size. In addition, the mean time to error occurrences is $(1/\lambda)/b_1 = 3600$–18000, the mean time until each task is executed again is $d_1/b_1 = 1$, the mean comparison time of the process of the job is $b_2/b_1 = 1$, the mean time until the job is executed again is $d_2/b_1 = 1$, and the probability that the comparison of the process results of the job matches is $q = 0.8$–1.0.

Table 3.3 presents the optimum number N^* that minimizes $\ell_{0E}(N)$ in (3.60). For example, when $a/b_1 = 200, (1/\lambda)/b_1 = 10800$, the optimum number is $N^* = 3$. This indicates that N^* decreases with $(1/\lambda)/b_1$, however,

Table 3.3. Optimum number N^* to minimize $\ell_{0E}(N)$

a/b_1	$(1/\lambda)/b_1$				
	3600	7200	10800	14400	18000
100	2	2	1	1	1
200	5	3	3	2	2
300	7	5	4	4	3
400	9	7	5	5	4

Table 3.4. Mean times $\ell_{0E}(1)$ and $\ell_{0E}(N^*)$

a/b_1	q	$(1/\lambda)/b_1$				
		3600	7200	10800	14400	18000
100	0.8	135	131	130	130	129
		133	131	130	130	129
	0.9	120	117	116	115	115
		118	116	116	115	115
	1.0	108	105	104	103	103
		106	104	104	103	103
200	0.8	281	267	262	260	258
		264	260	258	258	257
	0.9	249	237	233	231	230
		234	231	230	229	228
	1.0	224	213	209	208	206
		211	208	207	206	205
300	0.8	440	409	399	393	390
		395	389	387	386	384
	0.9	392	364	354	350	347
		351	346	344	342	342
	1.0	352	327	319	315	312
		315	311	309	308	307
400	0.8	614	559	540	531	525
		526	518	515	513	512
	0.9	546	496	480	472	467
		467	461	458	456	455
	1.0	491	447	432	424	420
		420	415	412	411	410

increases with a/b_1, *i.e.*, N^* increases as the job size becomes large. The partition number is at most nine under those parameters.

Table 3.4 presents the mean time $\ell_{0E}(N^*)$ when the job is partitioned into N^* tasks and $\ell_{0E}(1)$ when it is not done. This indicates that the mean time $\ell_{0E}(N^*)$ decreases with $(1/\lambda)/b_1$ and q. From the comparison with $\ell_{0E}(N^*)$ and $\ell_{0E}(1)$ in this table, the process time becomes about maximum 15% shorter by the partition with signature. In particular, the partition of the job into tasks is more effective in shortening one process time when a mean size a of a job is large. ∎

4

Maintenance Policies for a Finite Interval

It would be important to replace an operating unit before failure if its failure rate increases with age. The known results of replacement and maintenance policies were summarized [2]. After that, many papers and books were published and were extensively surveyed [1, 65–73]. Few papers treated maintenance for a finite time span because it is more difficult theoretically to discuss optimum policies for a finite time span. However, the working times of most units are finite in the actual field. For example, the number of old power plants in Japan is increasing, and civil structures and public infrastructures such as buildings, bridges, railroads, water supply, and drainage in advanced nations will become obsolete in the near future [74]. The importance of maintenance for aged units is much higher than that for new ones because the probabilities of occurrences of severe events would increase. Therefore, maintenance plans have to be reestablished at appropriate times for a specified finite interval.

An optimum sequential age replacement policy for a finite time span was derived [2] by using dynamic programming. A number of maintenance models with repair and replacement for finite horizons were reviewed [75]. The asymptotic costs of age replacement for a finite time span were given [76,77], and the finite inspection model with discounted costs when the failure time is exponential was discussed [78]. The inspection model for a finite working time is considered, and the optimum policy is given [48] by partitioning the working time into equal parts in Chap. 3. Maintenance policies for a finite interval were summarized [49]. It is of interest for such maintenance models that they can answer questions such as how many maintenance actions should be performed and when.

This chapter proposes the imperfect preventive maintenance (PM) model, where the failure rate increases with PM for a finite time span [79] in Sect. 4.1. The periodic policy in which the PM is done at ordered times kT $(k = 1, 2, \ldots, N)$, and the sequential policy in which it is done at times T_k $(k = 1, 2, \ldots, N)$ is considered. Next, we take up the periodic and sequential inspection policies in which a unit is checked at periodic or successive times for

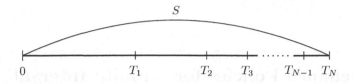

Fig. 4.1. Finite time S with sequential N intervals

a finite time span in Sect. 4.2. It is shown for both PM and inspection models how to compute optimum PM and inspection numbers and times numerically.

Finally, Section 4.3 applies periodic and sequential PM policies for a finite time span to a cumulative damage model [3]: The PM is done at periodic times kT and sequential times $\sum_{j=1}^{k} T_j$ $(k = 1, 2, \ldots, N)$ and reduces the total damage according to its improvement factor. When the total damage is x, a unit fails with probability $p(x)$ and undergoes a minimal repair at each failure. In particular, when shocks occur in a Poisson process and $p(x)$ is exponential, optimum PM times that minimize the expected cost are computed numerically. Such computations might be troublesome because we have to solve simultaneous equations with several variables, however, they would be easy as recent personal computers have developed greatly. Reliability properties and optimum maintenance policies for damage models are fully summarized [3].

4.1 Imperfect PM Policies

Several imperfect preventive maintenance (PM), where an operating unit becomes younger at each PM time, have been extensively summarized [80–83]. The unit becomes like new with a certain probability p, its age becomes x units younger, and its age t reduces to at after PM [1, p. 175]. Similar imperfect repair models were considered [82–84]. However, all models assumed that a unit is operating for an infinite time span. We apply such imperfect PM models to an operating unit which has to operate for a finite interval $[0, S]$.

Consider the following imperfect PM policy for an operating unit during $[0, S]$ [79]:

(1) The PM is done at planned times T_k $(k = 1, 2, \ldots, N - 1)$, and the unit is replaced at time T_N, where $T_N \equiv S$ and $T_0 \equiv 0$ (Fig. 4.1). The interval from T_{k-1} to T_k is called the kth PM period.
(2) The unit undergoes only minimal repair at failures during $[0, S]$, where minimal repair means that the failure rate remains undisturbed by any repair of failure [1, p. 96].

(3) The failure rate in the kth PM becomes $b_k h(x)$ when it was $h(x)$ in the $(k-1)$th PM, $i.e.$, the unit has the failure rate $B_k h(t)$ in the kth PM period for $0 < t \le T_k - T_{k-1}$, where $1 = b_0 < b_1 \le b_2 \le \cdots \le b_{N-1}$, $B_k \equiv \prod_{j=0}^{k-1} b_j$ $(k = 1, 2, \ldots, N)$, and $1 = B_1 < B_2 < \cdots < B_N$.

(4) The failure rate $h(t)$ increases strictly.

(5) The cost for each minimal repair is c_1, the cost for each PM is c_2, and the cost for replacement at time S is c_3.

(6) The times for PM, repair and replacement are negligible.

Because the expected number of minimal repairs during (T_{k-1}, T_k) is $\int_0^{T_k - T_{k-1}} B_k h(t) dt$ from assumption (3), the total expected cost until replacement is [79]

$$\mathbf{C}(N) = c_1 \sum_{k=1}^{N} B_k \int_0^{T_k - T_{k-1}} h(t)\, dt + (N-1)c_2 + c_3 \qquad (N = 1, 2, \ldots). \quad (4.1)$$

4.1.1 Periodic PM

Suppose that the PM is done at periodic times kT $(k = 1, 2, \ldots, N-1)$, and the unit is replaced at time NT, where $NT = S$ (Fig. 3.1). Then, the total expected cost is, from (4.1),

$$C(N) = c_1 \sum_{k=1}^{N} B_k H\left(\frac{S}{N}\right) + (N-1)c_2 + c_3, \qquad (4.2)$$

where $H(t) \equiv \int_0^t h(u)\, du$ represents the expected number of failures during $(0, t]$. In particular, when $B_k \equiv 1$ and $c_2 = c_3 = c_T$, $c_1 = c_M$, (4.2) agrees with (3.9). We find an optimum N^* that minimizes $C(N)$ in (4.2). Forming the inequality $C(N+1) - C(N) \ge 0$,

$$\sum_{k=1}^{N} B_k H\left(\frac{S}{N}\right) - \sum_{k=1}^{N+1} B_k H\left(\frac{S}{N+1}\right) \le \frac{c_2}{c_1} \qquad (N = 1, 2, \ldots). \quad (4.3)$$

When the failure time has a Weibull distribution, $i.e.$, $H(t) = \lambda t^m$ $(m > 1)$, (4.3) becomes

$$\left(\frac{1}{N}\right)^m \sum_{k=1}^{N} B_k - \left(\frac{1}{N+1}\right)^m \sum_{k=1}^{N+1} B_k \le \frac{c_2}{\lambda S^m c_1} \qquad (N = 1, 2, \ldots). \quad (4.4)$$

For example, if $\overline{B} \equiv \lim_{N \to \infty}(1/N)^m \sum_{k=1}^{N} B_k < \infty$, then the left-hand side of (4.4) goes to 0 as $N \to \infty$. In this case, a finite minimum number N^* $(1 \le N^* < \infty)$ to satisfy (4.3) always exists.

4.1.2 Sequential PM

We find optimum PM times T_k^* $(k = 1, 2, \ldots, N - 1)$ that minimize $\mathbf{C}(N)$ in (4.1). Differentiating $\mathbf{C}(N)$ with respect to T_k $(k = 1, 2, \ldots, N - 1)$ and setting it equal to zero,

$$h(T_k - T_{k-1}) = b_k h(T_{k+1} - T_k) \qquad (k = 1, 2, \ldots, N - 1). \qquad (4.5)$$

Noting that $T_0 = 0$, $T_N = S$, b_k (> 1) increases, and $h(t)$ increases strictly, optimum times T_k $(k = 1, 2, \ldots, N - 1)$ to satisfy (4.5) exist. For example, when $N = 3$, T_1 and T_2 are given by the solutions of the following equations:

$$h(T_1) = b_1 h(T_2 - T_1),$$
$$h(T_2 - T_1) = b_2 h(S - T_2).$$

From the above discussions, we compute T_k $(k = 1, 2, \ldots, N - 1)$ that satisfies (4.5), and substituting them in (4.1), we obtain the total expected cost $\mathbf{C}(N)$. Next, comparing $\mathbf{C}(N)$ for all $N \geq 1$, we can get the optimum PM number N^* and times T_k^* $(k = 1, 2, \ldots, N^*)$.

Example 4.1. Suppose that $H(t) = \lambda t^2$ and $S = 100$. In this case, we set the mean failure time equal to S, *i.e.*,

$$\int_0^\infty e^{-\lambda t^2} \, dt = \frac{1}{2}\sqrt{\frac{\pi}{\lambda}} \equiv S.$$

In addition, it is assumed that $b_k = 1 + k/(k+1)$ $(k = 0, 1, 2, \ldots)$ that increases strictly from 1 to 2. Then, $B_1 = 1$ and

$$B_{k+1} = \left(1 + \frac{1}{2}\right)\left(1 + \frac{2}{3}\right)\cdots\left(1 + \frac{k}{k+1}\right) \qquad (k = 1, 2, \ldots),$$

that increases strictly to ∞, and $\overline{B} = \lim_{N\to\infty}(1/N)^2 \sum_{k=1}^N B_k = 0$. Of course, if b_k is needed to increase from 1 to any number $n + 1$, then it may be assumed that $b_k = 1 + nk/(k+1)$.

Thus, from (4.3), an optimum N^* for periodic PM is given by a unique minimum that satisfies

$$\frac{2N + 1}{(N + 1)^2}\left(\frac{1}{N^2}\sum_{k=1}^N B_k - \frac{1}{N+1}B_N\right) \leq \frac{4c_2}{\pi c_1} \qquad (N = 1, 2, \ldots),$$

where note that its left-hand side decreases strictly from 3/8 to 0. Hence, if $4c_2/(\pi c_1) \geq 3/8$, *i.e.*, $c_1/c_2 < 3.4$ then $N^* = 1$, *i.e.*, no PM should be done until time S. Table 4.1 presents the optimum number N^* and the expected cost

$$\frac{\widetilde{C}(N^*)}{c_2} \equiv \frac{C(N^*) - c_3}{c_2} = \frac{\pi c_1}{4(N^*)^2 c_2}\sum_{k=1}^{N^*} B_k + N^* - 1$$

Table 4.1. Optimum number N^* and expected cost $\widetilde{C}(N^*)/c_2$ for periodic PM when $m = 2$ and $\lambda S^2 = \pi/4$

c_1/c_2	N^*	$\widetilde{C}(N^*)/c_2$
2	1	1.57
3	1	2.36
5	2	3.45
10	2	5.91
20	3	10.73
30	3	15.09

for $m = 2$ and $c_1/c_2 = 2, 3, 5, 10, 20$, and 30. Both N^* and $\widetilde{C}(N^*)$ increase gradually with c_1/c_2.

Next, optimum sequence times T_k $(k = 1, 2, \ldots, N - 1)$ from (4.5) when $H(t) = \lambda t^m$ $(m > 1)$ are given by solutions of the simultaneous equations:

$$\frac{T_1}{T_2 - T_1} = (b_1)^{1/(m-1)},$$

$$\frac{T_k - T_{k-1}}{T_{k+1} - T_k} = (b_k)^{1/(m-1)} \qquad (k = 2, 3, \ldots, N - 2),$$

$$\frac{T_{N-1} - T_{N-2}}{S - T_{N-1}} = (b_{N-1})^{1/(m-1)}.$$

Solving these equations,

$$T_k = \frac{S \sum_{j=1}^{k}[1/(B_j)^{1/(m-1)}]}{\sum_{j=1}^{N}[1/(B_j)^{1/(m-1)}]} \qquad (k = 1, 2, \ldots, N).$$

Table 4.2 presents the PM times T_k $(k = 1, 2, \ldots, N)$ and the expected cost

$$\frac{\widetilde{C}(N)}{c_2} \equiv \frac{C(N) - c_3}{c_2} = \frac{\pi c_1}{4S^2 c_2} \sum_{k=1}^{N} B_k (T_k - T_{k-1})^m + N - 1$$

for $m = 2$, $c_1/c_2 = 2, 3, 5, 10, 20$, and 30, and $S = 100$. This indicates that T_k decreases gradually with N and $\widetilde{C}(N)$ increases with c_1/c_2.

Comparing $\widetilde{C}(N)$ for $N = 1, 2, \ldots, 8$, the expected cost is minimum at $N^* = 1, 1, 2, 2, 3, 4$ for $c_1/c_2 = 2, 3, 5, 10, 20, 30$, respectively. The optimum N^* are the same as those for periodic PM except $c_1/c_2 = 30$. For example, when $S = 100$ and $c_1/c_2 = 30$, the PM should be done at times 43.57, 72.61, 90.04, and 100, and the expected cost is 13.266 and is $13.266/23.562 = 56.3\%$ and $13.266/15.09 = 87.9\%$ smaller than those of no PM and periodic PM, respectively. It is natural that the expected cost $\widetilde{C}(N)$ for sequential PM is equal to that when $N^* = 1$ and is smaller than that when $N^* \geq 2$ for periodic PM. ■

Table 4.2. Sequence PM times T_k and expected cost $\widetilde{C}(N)/c_2$ for $N = 1, 2, \ldots, 8$ when $S = 100$ and $m = 2$

N	1	2	3	4	5	6	7	8
T_1	100	60	48.39	43.57	41.28	40.14	39.54	39.24
T_2		100	80.65	72.61	68.81	66.89	65.91	65.39
T_3			100	90.04	85.32	82.95	81.73	81.09
T_4				100	94.76	92.12	90.76	90.06
T_5					100	97.22	95.79	95.04
T_6						100	98.53	97.76
T_7							100	99.22
T_8								100
c_1/c_2				$\widetilde{C}(N)/c_2$				
2	1.571	1.942	2.760	3.684	4.648	5.630	6.621	7.616
3	2.356	2.414	3.140	4.027	4.973	5.946	6.932	7.924
5	3.927	3.356	3.900	4.711	5.621	6.576	7.553	8.541
10	7.854	5.712	5.800	6.422	7.242	8.152	9.106	10.082
20	15.708	10.425	9.601	9.844	10.485	11.305	12.212	13.163
30	23.562	15.137	13.401	13.266	13.727	14.457	15.318	16.245

4.2 Inspection Policies

This section rewrites the standard inspection model for an infinite time span to the model in the finite case [1] and summarizes inspection policies for a finite interval $(0, S]$, in which its failure is detected only at inspection. Generally, it would be more troublesome to compute optimum inspection times in a finite case than those in an infinite case. In this section, we consider three inspection models of periodic, sequential, and asymptotic inspections.

In periodic inspection, an interval S is divided into N equal parts, and a unit is checked at periodic times kT $(k = 1, 2, \ldots, N - 1)$ and is replaced at time NT, where $NT \equiv S$. The optimum number N^* of checks has already been derived by using the partition method in Sect. 3.1.1. A numerical example is given when the failure time has a Weibull distribution. In sequential inspection, we show how to compute optimum checking times. In asymptotic inspection, we introduce an inspection intensity and show how to compute approximate checking times by a method simpler than that of the sequential one. Finally, we present numerical examples and show that the asymptotic inspection is a good approximation to sequential one. We can have similar discussions for obtaining an optimum policy that maximizes availability [85].

Table 4.3. Optimum number N^* and expected cost $\widetilde{C}(N^*)/c_2$ when $F(t) = 1 - e^{-\lambda t^2}$, $S = 100$ and $\lambda S^2 = \pi/4$

c_1/c_2	N^*	$\widetilde{C}(N^*)/c_2$
2	4	92.25
3	3	95.26
5	2	100.19
10	2	109.30
20	1	120.00
30	1	130.00

4.2.1 Periodic Inspection

A unit has to be operating for a finite interval $[0, S]$ and fails according to a general distribution $F(t)$. To detect failures, the unit is checked at periodic time kT ($k = 1, 2, \ldots, N$). Let c_1 be the cost for one check, c_2 be the cost per unit of time for the time elapsed between a failure and its detection at the next check, and c_3 be the replacement cost. Then, the total expected cost until failure detection or time S is [86]

$$C(N) = \sum_{k=0}^{N-1} \int_{kT}^{(k+1)T} \{c_1(k+1) + c_2[(k+1)T - t]\}\,dF(t) + c_1 N \overline{F}(NT) + c_3$$

$$= \left(c_1 + \frac{c_2 S}{N}\right) \sum_{k=0}^{N-1} \overline{F}\left(\frac{kS}{N}\right) - c_2 \int_0^S \overline{F}(t)\,dt + c_3 \qquad (N = 1, 2, \ldots).$$

$$(4.6)$$

It is evident that

$$C(1) = c_1 + c_2 \int_0^S \overline{F}(t)\,dt + c_3, \qquad C(\infty) \equiv \lim_{N \to \infty} C(N) = \infty.$$

Thus, there exists a finite number N^* ($1 \le N^* < \infty$) that minimizes $C(N)$. When $c_3 = 0$, the optimum policy that minimizes $C(N)$ is discussed in Sect. 3.1.1, using the partition method.

Example 4.2. Table 4.3 presents the optimum number N^* and the expected cost

$$\widetilde{C}(N^*) \equiv C(N^*) + c_2 \int_0^S \overline{F}(t)\,dt - c_3,$$

when $F(t) = 1 - e^{-\lambda t^2}$ and $S = 100$. In this case, we set the mean failure time equal to S as Example 4.1, *i.e.*, $\lambda S^2 = \pi/4$. The optimum N^* decreases as the check cost c_1 increases. ■

4.2.2 Sequential Inspection

An operating unit is checked at successive times $0 < T_1 < T_2 < \cdots < T_N$, where $T_0 \equiv 0$ and $T_N \equiv S$ (Fig. 4.1). Optimum inspection policies were surveyed [1, 68], and the finite inspection model with discounted cost when the failure time is exponential was discussed [78]. In a similar way for obtaining (4.6), the total expected cost until failure detection or time S is

$$\mathbf{C}(N) = \sum_{k=0}^{N-1} \int_{T_k}^{T_{k+1}} [c_1(k+1) + c_2(T_{k+1} - t)] \, dF(t) + c_1 N \overline{F}(T_N) + c_3$$

$$(N = 1, 2, \dots). \qquad (4.7)$$

Setting $\partial \mathbf{C}(N)/\partial T_k = 0$,

$$T_{k+1} - T_k = \frac{F(T_k) - F(T_{k-1})}{f(T_k)} - \frac{c_1}{c_2} \qquad (k = 1, 2, \dots, N-1), \qquad (4.8)$$

where $f(t)$ is a density function of $F(t)$, and the resulting expected cost is

$$\widetilde{\mathbf{C}}(N) \equiv \mathbf{C}(N) + c_2 \int_0^S \overline{F}(t) \, dt - c_3 = \sum_{k=0}^{N-1} [c_1 + c_2(T_{k+1} - T_k)] \overline{F}(T_k)$$

$$(N = 1, 2, \dots). \qquad (4.9)$$

From the above discussions, we compute T_k $(k = 1, 2, \dots, N-1)$ that satisfies (4.8), and substituting them in (4.9), we obtain the expected cost $\mathbf{C}(N)$. Note that (4.8) agrees with (3.5) of [2, p. 110], and (4.7) agrees with (8.1) of [1, p. 203] as $N \to \infty$ when $c_3 = 0$. Next, comparing $\mathbf{C}(N)$ for all $N \geq 1$, we can get the optimum checking number N^* and times T_k^* $(k = 1, 2, \dots, N^*)$.

Example 4.3. Table 4.4 presents the checking time T_k $(k = 1, 2, \dots, N)$ and the expected cost $\widetilde{\mathbf{C}}(N)/c_2$ for $S = 100$ and $c_1/c_2 = 2$ when $F(t) = 1 - e^{-\lambda t^2}$ and $\lambda S^2 = \pi/4$. Comparing $\widetilde{\mathbf{C}}(N)$ for $N = 1, 2, \dots, 8$, the expected cost is minimum at $N^* = 4$. In this case, the optimum number is the same, however, the expected cost is $91.16/92.25 = 98.8\%$ smaller compared with that in Table 4.3 when $c_1/c_2 = 2$.

Next, consider the problem of minimizing the total expected cost $\mathbf{C}(N)$ under a constraint of the inspection cost [87]. We compute an optimum number N^* subject to $c_1 \widetilde{N} \leq C$, i.e., $\widetilde{N} \equiv [C/c_1]$. Then, the optimum policy is derived from Table 4.4 as follows: If $N^* \leq \widetilde{N}$, then the optimum number is N^*, and if $N^* > \widetilde{N}$, then it is $\widetilde{N}$. ∎

4.2.3 Asymptotic Inspection

Define that $n(t)$ is a smooth inspection intensity, i.e., $n(t)dt$ is the probability that a unit is checked for a small interval $(t, t+dt)$ [88]. Then, the approximate total expected cost [88, 89] is given by

Table 4.4. Checking time T_k and expected cost $\widetilde{C}(N)/c_2$ when $S = 100$, $c_1/c_2 = 2$ and $F(t) = 1 - e^{-\lambda t^2}$

N	1	2	3	4	5	6	7	8
T_1	100	64.14	50.9	44.1	40.3	38.1	36.8	36.3
T_2		100	77.1	66.0	60.0	56.2	54.3	53.3
T_3			100	84.0	75.4	70.5	67.8	66.6
T_4				100	88.6	82.3	78.9	77.3
T_5					100	91.1	87.9	85.9
T_6						100	94.9	92.5
T_7							100	97.2
T_8								100
$\widetilde{C}(N)/c_2$	102.00	93.55	91.52	91.16	91.47	92.11	92.91	93.79

$$C(n(t)) = \int_0^S \left[c_1 \int_0^t n(x)\,dx + \frac{c_2}{2n(t)} \right] dF(t) + c_1 \overline{F}(S) \int_0^S n(t)\,dt + c_3.$$

$$(4.10)$$

Letting $h(t)$ be the failure rate of $F(t)$, differentiating $C(n(t))$ with respect to $n(t)$, and setting it equal to zero,

$$n(t) = \sqrt{\frac{h(t)c_2}{2c_1}}.$$

$$(4.11)$$

An inspection density $n(t)$ was also given by solving the Euler equation in (4.10) [90].

We compute approximate checking times $\widetilde{T}_k$ ($k = 1, 2, \ldots, N-1$) and a checking number $\widetilde{N}$ by using (4.11). First, we set

$$\int_0^S \sqrt{\frac{h(t)c_2}{2c_1}}\,dt \equiv X$$

and $[X] \equiv N$, where $[x]$ denotes the greatest integer contained in x. Then, we obtain A_N ($0 < A_N \leq 1$) such that

$$A_N \int_0^S \sqrt{\frac{h(t)c_2}{2c_1}}\,dt = N,$$

and define an inspection intensity as

$$\tilde{n}(t) = A_N \sqrt{\frac{h(t)c_2}{2c_1}}.$$

$$(4.12)$$

Table 4.5. Checking time T_k and expected cost $\widetilde{C}(N)/c_2$ for $N = 4, 5$ when $S = 100$, $c_1/c_2 = 2$ and $F(t) = 1 - e^{-\lambda t^2}$

N	4	5
1	39.7	34.2
2	63.0	54.3
3	82.5	71.1
4	100.0	86.2
5		100.0
$\widetilde{C}(N)/c_2$	91.22	91.58

Using (4.12), we compute checking times T_k that satisfy

$$\int_0^{T_k} \widetilde{n}(t)\, dt = k \qquad (k = 1, 2, \ldots, N-1), \tag{4.13}$$

where note that $T_0 = 0$ and $T_N = S$. Then, the total expected cost is given in (4.7).

Next, we set N by $N+1$ and do similar computations. At last, we compare $\mathbf{C}(N)$ and $\mathbf{C}(N+1)$, and choose the smallest as the total expected cost $\mathbf{C}(\widetilde{N})$ and checking times $\widetilde{T}_k$ ($k = 1, 2, \ldots, \widetilde{N}$) as an asymptotic inspection policy.

Example 4.4. Consider a numerical example when the parameters are the same as those of Example 4.3, *i.e.*, $S = 100$, $c_1/c_2 = 2$ and $F(t) = 1 - e^{-\lambda t^2}$. Then, because $\lambda = \pi/(4 \times 10^4)$, $n(t) = \sqrt{\lambda t/2}$, $[X] = N = 4$, and $A_N = (12/100)/\sqrt{\pi/200}$, $\widetilde{n}(t) = 6\sqrt{t}/10^3$. Thus, from (4.13), checking times are

$$\int_0^{T_k} \frac{6}{1000} \sqrt{t}\, dt = \frac{1}{250} T_k^{3/2} = k \qquad (k = 1, 2, 3).$$

When $N = 5$, $A_N = (15/100)/\sqrt{\pi/200}$ and $\widetilde{n}(t) = 3\sqrt{t}/(4 \times 10^2)$. In this case, checking times are

$$\int_0^{T_k} \frac{3}{400} \sqrt{t}\, dt = \frac{1}{200} T_k^{3/2} = k \qquad (k = 1, 2, 3, 4).$$

Table 4.5 presents the checking times T_k and the resulting expected costs $\widetilde{C}(N)/c_2$ for $N = 4$ and 5. Because $\widetilde{C}(4) < \widetilde{C}(5)$, the approximate checking number is $\widetilde{N} = 4$, and its checking times $\widetilde{T}_k$ are 39.7, 63.0, 82.5, and 100. These checking times are a little smaller than those in Table 4.4 when $N = 4$, however, they closely approximate the optimum ones. Furthermore, the expected cost 91.22 is a little greater than 91.16, however, it is smaller than 92.25 in Table 4.3 for periodic inspection. ∎

4.3 Cumulative Damage Models

Many serious accidents have happened recently and caused heavy damage as systems have become large-scale and complex. Everyone is very anxious that big earthquakes might happen in the near future in Japan and might destroy large high buildings and old chemical and power plants, and inflict serious damage on wide areas. Furthermore, public infrastructure in most advanced nations is becoming old. Maintenance policies for such industrial systems and public infrastructure should be established scientifically and practically according to their circumstances [1].

As one example of maintenance models, we can consider the cumulative damage model where the total damage is additive. Such reliability models and their optimum maintenance policies were discussed extensively [3]. We can apply the cumulative damage model to the gas turbine engine of a cogeneration system [22]: A gas turbine engine is adopted mainly as the power source of a cogeneration system because it is small, its exhaust gas emission is clean, and both its noise and vibration level are low. The turbine engine suffers mechanical damage when it is turned on and operated, so that the engine has to be overhauled when it has exceeded the number of turning-on or the total operating time. Damage models were also applied to crack growth models [91–93], welded joints [94], floating structures [95], reinforced concrete structures [96], and plastic automotive components [97]. Such stochastic models of fatigue damage of materials in engineering systems were described in detail [98–100].

We take up the cumulative damage model with minimal repair at failure [101]: The unit is subject to shocks that occur in a Poisson process. At each shock, the unit suffers random damage which is additive and fails with probability $p(x)$ when the total damage is x. If the unit fails, then it undergoes only minimal repair. We apply a sequential PM policy to this model where each PM is imperfect: The PM is done at the intervals of sequential times T_k, and the unit is replaced at time S. Note carefully in this section that T_k is denoted by the intervals between PMs for the simplicity of equations. The amount of damage after the kth PM becomes $a_k Z_k$ when it was Z_k before PM, i.e., the kth PM reduces the total damage Z_k to $a_k Z_k$. Furthermore, suppose that the unit has to be operating for a finite interval $(0, S]$. Then, setting $\sum_{k=1}^{N} T_k = S$, we compute an optimum number N^* and optimum times T_k^* ($k = 1, 2, \ldots, N^* - 1$) that minimize the expected cost until replacement.

Consider a sequential PM policy for the unit, where the PM is done at fixed intervals T_k ($k = 1, 2, \ldots, N - 1$), and the replacement is done at $T_1 + T_2 + \cdots + T_N = S$ [102] (Fig. 4.2). We call an interval from the $(k - 1)$th PM to the kth PM *period k*.

Suppose that shocks occur in a Poisson process with rate λ. Random variables Y_k ($k = 1, 2, \ldots, N$) denote the number of shocks in period k, i.e., $\Pr\{Y_k = j\} = [(\lambda T_k)^j / j!] \exp(-\lambda T_k)$ ($j = 0, 1, 2, \ldots$). In addition, we denote W_{kj} the amount of damage caused by the jth shock in period k,

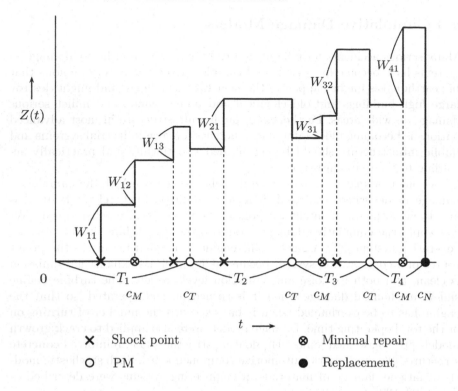

Fig. 4.2. Process for Imperfect PM with PM intervals T_k

where $W_{k0} \equiv 0$. It is assumed that W_{kj} are nonnegative, independent, and identically distributed random variables and have an identical distribution $\Pr\{W_{kj} \leq x\} \equiv G(x)$ for all k and j. The total damage is additive, and $G^{(j)}(x)$ $(j = 1, 2, \dots)$ is the j-fold Stieltjes convolution of $G(x)$ with itself and $G^{(0)}(x) \equiv 1$ for all $x \geq 0$. Then, it follows that

$$\Pr\{W_{k1} + W_{k2} + \cdots + W_{kj} \leq x\} = G^{(j)}(x) \qquad (j = 0, 1, 2, \dots). \qquad (4.14)$$

When the total damage becomes x at some shock, the unit fails with probability $p(x)$ that increases with x from 0 to 1. If the unit fails between PMs, it undergoes only minimal repair, and hence, the total damage remains unchanged by any minimal repair. It is assumed that all times required for any PM and minimal repair are negligible.

Next, we introduce an improvement factor in PM, where the kth PM reduces $100(1 - a_k)\%$ $(0 \leq a_k \leq 1)$ of the total damage. Letting Z_k be the total damage at the end of period k, i.e., just before the kth PM, the kth PM reduces it to $a_k Z_k$. Because the total damage during period k is additive and is not removed by any minimal repair,

$$Z_k = a_{k-1}Z_{k-1} + \sum_{j=1}^{Y_k} W_{kj} \qquad (k = 1, 2, \ldots, N), \qquad (4.15)$$

where $Z_0 \equiv 0$ and $\sum_{j=1}^{0} \equiv 0$.

Let c_T be the cost for each PM, c_N be the cost for replacement with $c_N > c_T$, and c_M be the cost for minimal repair. Then, from the assumption that the unit fails with probability $p(\cdot)$ only at shocks, the total cost in period k is

$$\widetilde{C}(k) = c_T + c_M \sum_{j=1}^{Y_k} p(a_{k-1}Z_{k-1} + W_{k1} + W_{k2} + \cdots + W_{kj})$$

$$(k = 1, 2, \ldots, N-1), \qquad (4.16)$$

$$\widetilde{C}(N) = c_N + c_M \sum_{j=1}^{Y_N} p(a_{N-1}Z_{N-1} + W_{N1} + W_{N2} + \cdots + W_{Nj}). \qquad (4.17)$$

Furthermore, we assume that $p(x)$ is exponential, i.e., $p(x) = 1 - e^{-\theta x}$ for $\theta > 0$. Letting $G^*(\theta)$ be the Laplace-Stieltjes transform of $G(x)$, i.e., $G^*(\theta) \equiv \int_0^\infty e^{-\theta x} dG(x)$,

$$E\{\exp[-\theta(W_{k1} + W_{k2} + \cdots + W_{kj})]\} = \int_0^\infty e^{-\theta x} dG^{(j)}(x) = [G^*(\theta)]^j. \quad (4.18)$$

Using the law of total probability in (4.16), the expected cost in period k is

$$E\{\widetilde{C}(k)\} = c_T + c_M E \left\{ \sum_{j=1}^{Y_k} p(a_{k-1}Z_{k-1} + W_{k1} + W_{k2} + \cdots + W_{kj})] \right\}$$

$$= c_T + c_M \sum_{i=1}^{\infty} \sum_{j=1}^{i} E\{1 - \exp[-\theta(a_{k-1}Z_{k-1} + W_{k1} + W_{k2} + \cdots + W_{kj})]\} \Pr\{Y_k = i\}.$$

Let $B_k^*(\theta) \equiv E\{\exp(-\theta Z_k)\}$. Then, because Z_{k-1} and W_{kj} are independent of each other, from (4.18),

$$E\{1 - \exp[-\theta(a_{k-1}Z_{k-1} + W_{k1} + \cdots + W_{kj})]\} = 1 - B_{k-1}^*(\theta a_{k-1})[G^*(\theta)]^j.$$

Thus, from the assumption that Y_k has a Poisson distribution with rate λ,

$$E\{\widetilde{C}(k)\} = c_T + c_M \sum_{i=1}^{\infty} \frac{(\lambda T_k)^i}{i!} e^{-\lambda T_k} \sum_{j=1}^{i} \{1 - B_{k-1}^*(\theta a_{k-1})[G^*(\theta)]^j\}$$

$$= c_T + c_M \left\{ \lambda T_k - \frac{G^*(\theta)}{1 - G^*(\theta)} B_{k-1}^*(\theta a_{k-1})[1 - e^{-\lambda[1-G^*(\theta)]T_k}] \right\}$$

$$(k = 1, 2, \ldots, N-1). \qquad (4.19)$$

Similarly, the expected cost in period N is

$$E\{\widetilde{C}(N)\} = c_N + c_M \left\{ \lambda T_N - \frac{G^*(\theta)}{1 - G^*(\theta)} B^*_{N-1}(\theta a_{N-1})[1 - e^{-\lambda[1 - G^*(\theta)]T_N}] \right\}. \tag{4.20}$$

Letting $A_r^k \equiv \prod_{j=r}^k a_j$ for $r \le k$ and $\equiv 0$ for $r > k$, from (4.15),

$$a_{k-1} Z_{k-1} = \sum_{r=1}^{k-1} A_r^{k-1} \sum_{j=1}^{Y_r} W_{rj}.$$

Thus, recalling that W_{ij} are independent and have an identical distribution $G(x)$ [101],

$$B_{k-1}(\theta a_{k-1}) = E\{e^{-\theta a_{k-1} Z_{k-1}}\} = E\left\{ \exp\left[-\theta \sum_{r=1}^{k-1} A_r^{k-1} \sum_{j=1}^{Y_r} W_{rj} \right] \right\}.$$

Because

$$E\left\{ \exp\left[-\theta A_r^{k-1} \sum_{j=1}^{Y_r} W_{rj} \right] \right\} = \sum_{i=0}^{\infty} \Pr\{Y_r = i\} E\left\{ \exp\left[-\theta A_r^{k-1} \sum_{j=1}^{i} W_{rj} \right] \right\}$$

$$= \sum_{i=0}^{\infty} \frac{(\lambda T_r)^i}{i!} e^{-\lambda T_r} [G^*(\theta A_r^{k-1})]^i$$

$$= \exp\{-\lambda T_r [1 - G^*(\theta A_r^{k-1})]\},$$

consequently,

$$B_{k-1}(\theta a_{k-1}) = \exp\left\{ -\sum_{j=1}^{k-1} \lambda T_j [1 - G^*(\theta A_j^{k-1})] \right\}. \tag{4.21}$$

Substituting (4.21) in (4.19) and (4.20), respectively, the expected cost in period k is

$$E\{\widetilde{C}(k)\} = c_T + c_M \left(\lambda T_k - \frac{G^*(\theta)}{1 - G^*(\theta)} \exp\left\{ -\sum_{j=1}^{k-1} \lambda T_j [1 - G^*(\theta A_j^{k-1})] \right\} \right.$$

$$\left. \times \left\{ 1 - e^{-\lambda T_k [1 - G^*(\theta)]} \right\} \right) \quad (k = 1, 2, \ldots, N-1), \tag{4.22}$$

$$E\{\widetilde{C}(N)\} = c_N + c_M \left(\lambda T_N - \frac{G^*(\theta)}{1 - G^*(\theta)} \exp\left\{ -\sum_{j=1}^{N-1} \lambda T_j [1 - G^*(\theta A_j^{N-1})] \right\} \right.$$

$$\left. \times \left\{ 1 - e^{-\lambda T_N [1 - G^*(\theta)]} \right\} \right). \tag{4.23}$$

Therefore, the total expected cost until replacement is

$$\mathbf{C}(N) = \sum_{k=1}^{N-1} E\{\widetilde{C}(k)\} + E\{\widetilde{C}(N)\}$$

$$= (N-1)c_T + c_N + c_M \lambda S$$

$$- c_M \frac{G^*(\theta)}{1-G^*(\theta)} \sum_{k=1}^{N} \exp\left\{-\sum_{j=1}^{k-1} \lambda T_j [1 - G^*(\theta A_j^{k-1})]\right\} \left\{1 - e^{-\lambda T_k [1 - G^*(\theta)]}\right\}$$

$$(N = 1, 2, \dots). \quad (4.24)$$

4.3.1 Periodic PM

Suppose that the PM is done at periodic times kT $(k = 1, 2, \dots, N-1)$, and the unit is replaced at time NT, where $NT = S$. It is assumed that $a_k \equiv a$ and $G(x) = 1 - e^{-\mu x}$. Then, the total expected cost in (4.24) is rewritten as

$$\widetilde{\mathbf{C}}(N) \equiv c_M \lambda S - \mathbf{C}(N)$$

$$= \frac{\mu c_M}{\theta} \left\{1 - e^{-(\lambda S/N)[\theta/(\theta+\mu)]}\right\} \sum_{k=1}^{N} Q(k|N) - (N-1)c_T - c_N$$

$$(N = 1, 2, \dots), \quad (4.25)$$

where

$$Q(k|N) \equiv \exp\left[-\frac{\lambda S}{N} \sum_{j=1}^{k-1} \left(\frac{\theta a^{k-j}}{\theta a^{k-j} + \mu}\right)\right].$$

We find an optimum number N^* that maximizes $\widetilde{\mathbf{C}}(N)$. Forming the inequality $\widetilde{\mathbf{C}}(N+1) - \widetilde{\mathbf{C}}(N) \leq 0$,

$$\left[1 - \exp\left(-\frac{\lambda S}{N+1} \frac{\theta}{\theta+\mu}\right)\right] \sum_{k=1}^{N+1} Q(k|N+1) - \left[1 - \exp\left(-\frac{\lambda S}{N} \frac{\theta}{\theta+\mu}\right)\right] \sum_{k=1}^{N} Q(k|N)$$

$$\leq \frac{\theta c_T}{\mu c_M} \quad (N = 1, 2, \dots). \quad (4.26)$$

Thus, if the left-hand side of (4.26) decreases strictly with N, then there exists a unique minimum N^* that satisfies (4.26).

Example 4.5. Table 4.6 presents the optimum number N^* and $\widetilde{\mathbf{C}}(N^*)/c_M$ for c_T/c_M. This indicates that the optimum N^* decreases with c_T/c_M because the cost for PM increases compared with the cost for minimal repair. ∎

Table 4.6. Optimum number N^* and expected cost $\widetilde{C}(N^*)/c_M$ when $a = 0.5$, $\mu/\theta = 10$, $c_T/c_M = 5$ and $\lambda S = 40$

c_T/c_M	N^*	$\widetilde{C}(N^*)/c_M$
0.1	41	22.814
0.2	27	19.608
0.5	15	13.893
1.0	9	8.637
1.5	5	5.739
2.0	1	4.737

4.3.2 Sequential PM

Suppose that the PM is done at sequential interval times T_k ($k = 1, 2, \ldots, N-1$), and the unit is replaced at time S, where $\sum_{k=1}^{N} T_k = S$. When $a_k = a$ and $G(x) = 1 - e^{-\mu x}$, the total expected cost in (4.24) is rewritten as

$$
\widetilde{C}(N) \equiv c_M \lambda S - \mathbf{C}(N)
$$

$$
= \frac{\mu c_M}{\theta} \sum_{k=1}^{N} \exp\left[-\sum_{j=1}^{k-1} \lambda T_j \left(\frac{\theta a^{k-j}}{\theta a^{k-j} + \mu}\right)\right]
$$

$$
\times \left(1 - e^{-\lambda T_k \frac{\theta}{\theta+\mu}}\right) - (N-1)c_T - c_N \qquad (N = 1, 2, \ldots). \quad (4.27)
$$

For example, when $N = 1$,

$$
\widetilde{C}(S) = \frac{\mu c_M}{\theta}\left(1 - e^{-\lambda S \frac{\theta}{\theta+\mu}}\right) - c_N. \quad (4.28)
$$

When $N = 2$,

$$
\widetilde{C}(T_1) = \frac{\mu c_M}{\theta}\left\{1 - e^{-\lambda T_1 \frac{\theta}{\theta+\mu}} + e^{-\lambda T_1 \frac{\theta a}{\theta a+\mu}}\left[1 - e^{-\lambda(S-T_1)\frac{\theta}{\theta+\mu}}\right]\right\} - c_T - c_N. \quad (4.29)
$$

Differentiating $\widetilde{C}(T_1)$ with respect to T_1 and setting it equal to zero,

$$
\frac{\theta}{\theta+\mu}\left[e^{-\lambda T_1\left(\frac{\theta}{\theta+\mu} - \frac{\theta a}{\theta a+\mu}\right)} - e^{-\lambda(S-T_1)\frac{\theta}{\theta+\mu}}\right] - \frac{\theta a}{\theta a+\mu}\left[1 - e^{-\lambda(S-T_1)\frac{\theta}{\theta+\mu}}\right] = 0. \quad (4.30)
$$

Letting $Q(T)$ be the left-hand side of (4.30),

$$
Q(0) = \left(\frac{\theta}{\theta+\mu} - \frac{\theta a}{\theta a+\mu}\right)\left(1 - e^{-\lambda S \frac{\theta}{\theta+\mu}}\right) > 0,
$$

$$Q(S) = -\frac{\theta}{\theta+\mu}\left[1 - e^{-\lambda S\left(\frac{\theta}{\theta+\mu} - \frac{\theta a}{\theta a+\mu}\right)}\right] < 0,$$

$$Q'(T) = -\left(\frac{\theta}{\theta+\mu}\right)\left(\frac{\theta}{\theta+\mu} - \frac{\theta a}{\theta a+\mu}\right)\left[e^{-\lambda T_1\left(\frac{\theta}{\theta+\mu} - \frac{\theta a}{\theta a+\mu}\right)} + e^{-\lambda(S-T_1)\frac{\theta}{\theta+\mu}}\right] < 0.$$

Thus, there exists an optimum T_1^* ($0 < T_1^* < S$) that satisfies (4.30).

When $N = 3$,

$$\widetilde{C}(T_1, T_2) = \frac{\mu c_M}{\theta}\left\{1 - e^{-\lambda T_1\frac{\theta}{\theta+\mu}} + e^{-\lambda T_1\frac{\theta a}{\theta a+\mu}}\left(1 - e^{-\lambda T_2\frac{\theta}{\theta+\mu}}\right)\right.$$
$$\left. + e^{-\lambda T_1\frac{\theta a^2}{\theta a^2+\mu} - \lambda T_2\frac{\theta a}{\theta a+\mu}}\left[1 - e^{-\lambda(S-T_1-T_2)\frac{\theta}{\theta+\mu}}\right]\right\} - 2c_T - c_N. \quad (4.31)$$

Differentiating $\widetilde{C}(T_1, T_2)$ with respect to T_1 and T_2 and setting them equal to zero, respectively,

$$\frac{\theta}{\theta+\mu}\left[e^{-\lambda T_1\frac{\theta}{\theta+\mu}} - e^{-\lambda T_1\frac{\theta a^2}{\theta a^2+\mu} - \lambda T_2\frac{\theta a}{\theta a+\mu} - \lambda(S-T_1-T_2)\frac{\theta}{\theta+\mu}}\right]$$
$$- \frac{\theta a}{\theta a+\mu}e^{-\lambda T_1\frac{\theta a}{\theta a+\mu}}\left(1 - e^{-\lambda T_2\frac{\theta}{\theta+\mu}}\right)$$
$$- \frac{\theta a^2}{\theta a^2+\mu}e^{-\lambda T_1\frac{\theta a^2}{\theta a^2+\mu} - \lambda T_2\frac{\theta a}{\theta a+\mu}}\left[1 - e^{-\lambda(S-T_1-T_2)\frac{\theta}{\theta+\mu}}\right] = 0, \quad (4.32)$$

$$\frac{\theta}{\theta+\mu}\left[e^{-\lambda T_1\frac{\theta a}{\theta a+\mu} - \lambda T_2\frac{\theta}{\theta+\mu}} - e^{-\lambda T_1\frac{\theta a^2}{\theta a^2+\mu} - \lambda T_2\frac{\theta a}{\theta a+\mu} - \lambda(S-T_1-T_2)\frac{\theta}{\theta+\mu}}\right]$$
$$- \frac{\theta a}{\theta a+\mu}e^{-\lambda T_1\frac{\theta a^2}{\theta a^2+\mu} - \lambda T_2\frac{\theta a}{\theta a+\mu}}\left[1 - e^{-\lambda(S-T_1-T_2)\frac{\theta}{\theta+\mu}}\right] = 0. \quad (4.33)$$

In general, differentiating $\widetilde{C}(N)$ with respect to T_k ($k = 1, 2, \ldots, N-1$) ($N \geq 2$) and setting them equal to zero,

$$\frac{\theta}{\theta+\mu}\left[\exp\left(-\sum_{j=1}^{k}\lambda T_j\frac{\theta a^{k-j}}{\theta a^{k-j}+\mu}\right) - \exp\left(-\sum_{j=1}^{N}\lambda T_j\frac{\theta a^{n-j}}{\theta a^{n-j}+\mu}\right)\right]$$
$$- \sum_{i=k+1}^{N}\frac{\theta a^{i-k}}{\theta a^{i-k}+\mu}\exp\left(-\sum_{j=1}^{i-1}\lambda T_j\frac{\theta a^{i-j}}{\theta a^{i-j}+\mu}\right)\left(1 - e^{-\lambda T_i\frac{\theta}{\theta+\mu}}\right) = 0$$

$$(k = 1, 2, \ldots, N-1), \quad (4.34)$$

where note that $T_N = S - T_1 - T_2 - \cdots - T_{N-1}$. Therefore, we may solve the simultaneous equations (4.34) and obtain the expected cost $\widetilde{C}(N)$ in (4.27). Next, comparing $\widetilde{C}(N)$ for all $N \geq 1$, we can get the optimum number N^* and times T_k^* ($k = 1, 2, \ldots, N^*-1$) for a specified S.

Table 4.7. PM time interval λT_k and expected cost $\widetilde{C}(N)/c_M$ when $a = 0.5$, $\mu/\theta = 10$, $c_N/c_M = 5$, $c_T/c_M = 1.0$, and $\lambda S = 40$

N	1	2	3	4	5	6	7	8	9	10
λT_1	40.00	13.17	12.41	11.37	10.32	9.36	8.52	7.80	7.17	6.63
λT_2		26.83	5.60	5.27	4.82	4.38	3.99	3.66	3.37	3.11
λT_3			21.99	5.23	4.87	4.45	4.06	3.72	3.42	3.17
λT_4				18.22	4.78	4.45	4.07	3.73	3.44	3.18
λT_5					15.22	4.35	4.06	3.73	3.44	3.18
λT_6						13.01	3.97	3.71	3.44	3.18
λT_7							11.33	3.64	3.42	3.18
λT_8								10.01	3.35	3.16
λT_9									8.96	3.10
λT_{10}										8.10
$\widetilde{C}(N)/c_M$	4.74	5.86	6.87	7.70	8.34	8.78	9.05	9.17	9.16	9.03

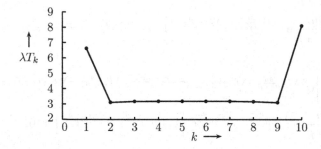

Fig. 4.3. Optimum PM time interval λT_k for $k = 1, 2, \ldots, 10$

Example 4.6. Table 4.7 presents λT_k and $\widetilde{C}(N)/c_M$ when $a = 0.5$, $\mu/\theta = 10$, $c_N/c_M = 5$, $c_T/c_M = 1.0$, and $\lambda S = 40$. Comparing $\widetilde{C}(N)$ for $N = 1, 2, \ldots$, 10, the expected cost $\widetilde{C}(N)$ is maximum, *i.e.*, $C(N)$ in (4.24) is minimum at $N = 8$. In this case, the optimum PM number is $N^* = 8$ and $N^* = 9$ for periodic PM, and $\widetilde{C}(8)/c_M$ is greater than 8.637 in Table 4.6 for periodic PM. The optimum PM times are 7.80, 11.46, 15.18, 18.91, 22.64, 26.35, 29.99, and 40 when $\lambda = 1$. This indicates that the last PM time interval is the largest and the first one is the second, and they increase first, remain constant for some number, and then decrease for large N, *i.e.*, the PM time intervals draw an upside-down bathtub curve [103] for $2 \leq k \leq N - 1$. Figure 4.3 shows the PM interval times T_k ($k = 1, 2, \ldots, 10$) and draws roughly a standard bathtub curve. ■

5

Forward and Backward Times in Reliability Models

The most important problem in reliability theory is to estimate statistically at what time an operating unit will fail in the near future. From such reliability viewpoints, failure distributions and their parameters have been estimated statistically, and some reliability quantities have been well defined and obtained. Systems with high reliability have been designed, and maintenance policies to prevent failures have been discussed analytically, collecting a large amount of information on failure times of object units. We call such times in the future *forward times*. Most problems in reliability are to solve practical problems concerning forward times. Reliability theory has been developed greatly through probabilistic investigation on forward times [1, 2, 73, 104].

On the other hand, when a unit is detected to have failed, and its failure time is unknown, we often want to know the past time when it failed. We call the time that goes back from failure detection to failure time *backward times*. There exist some optimization problems of backward times in actual reliability models. For example, suppose that some products are weighed and shipped out, using a scale whose accuracy is checked every day. Then, one problem is how much product we have to reweigh when the scale is uncalibrated and is judged to be inaccurate [105]. Another example is the backup policy for a database system [106]: When a failure has occurred in the process of a database system, we execute the rollback operation until the latest checkpoint and make it the recovery. The problem is when to place ordered checkpoints at planned times.

In this chapter, we summarize the properties of forward and backward times, using the failure rate and the reversed failure rate. It is of great interest that two times have symmetrical properties. As applied problems with forward times, we propose modified age replacement models. Furthermore, we take up the work of a job that has a scheduling time and is achieved by a unit, and discuss analytically an optimum scheduling time [107]. For backward times, we consider an optimization problem of how much time we go back to detect failure [108] and attempt to apply the backward model to the traceability problem in production systems. As one traceability policy, we discuss analytically whether or not the record of a production should be kept. Furthermore, as practical applications, we take up the backup model of a database system [109] and the model of reweighing by a scale [105]. There exist a number of actual models where we go back to some point and restore a normal condition after maintenance, when a failure has been detected.

5.1 Forward Time

Suppose that a unit begins to operate at time 0, and a random variable X denotes its failure time with a probability distribution $F(t) \equiv \Pr\{X \leq t\}$, a density function $f(t)$, and its finite mean $\mu \equiv E\{X\} = \int_0^\infty t f(t) \mathrm{d}t = \int_0^\infty \overline{F}(t) \mathrm{d}t < \infty$, where $\overline{F}(t) \equiv 1 - F(t)$. Then, the probability that a unit fails during $(t, t+x]$ $(0 \leq t < \infty)$, given that it has not failed in time t (Fig. 5.1), is

$$F(x|t) \equiv \Pr\{t < X \leq t + x | X > t\} = \frac{F(t+x) - F(t)}{\overline{F}(t)}, \tag{5.1}$$

$$\overline{F}(x|t) \equiv 1 - F(x|t) = \frac{\overline{F}(t+x)}{\overline{F}(t)} \tag{5.2}$$

for $0 \leq x < \infty$ and $F(t) < 1$, and its density function is

$$f(x|t) \equiv \frac{\mathrm{d}F(x|t)}{\mathrm{d}x} = \frac{f(t+x)}{\overline{F}(t)}. \tag{5.3}$$

Thus, the mean residual time from time t to failure is

$$\alpha(t) \equiv \int_0^\infty \overline{F}(x|t) \, \mathrm{d}x = \frac{1}{\overline{F}(t)} \int_t^\infty \overline{F}(x) \, \mathrm{d}x. \tag{5.4}$$

We summarize briefly the main properties of $F(x|t)$, $f(x|t)$, and $\alpha(t)$ [108]:

(1) When $t = 0$, $F(x|0) = F(x)$, $F(0|t) = 0$, $F(\infty|t) = 1$, $\alpha(0) = \mu$, and $f(0|t) = f(t)/\overline{F}(t) \equiv h(t)$. Thus,

$$\overline{F}(x|t) = \mathrm{e}^{-\int_t^{t+x} h(u) \, \mathrm{d}u}. \tag{5.5}$$

Note that both $h(t)$ and $F(x|t)$ are called the *failure rate* or *hazard rate* and have the same properties [2] because

$$h(t) = \frac{1}{\overline{F}(t)} \lim_{x \to 0} \frac{F(t+x) - F(t)}{x}.$$

(2) When $F(t) = 1 - \exp(-\lambda t^m)$ $(m > 0)$, $F(x|t) = 1 - \exp[-\lambda(t+x)^m - \lambda t^m]$ and $h(t) = \lambda m t^{m-1}$. Thus, the failure rate $h(t)$ decreases strictly for $0 < m < 1$, is constant for $m = 1$, and increases strictly for $m > 1$. Furthermore, $F(x) = F(x|t) = 1 - \mathrm{e}^{-\lambda x}$, and $\alpha(t) = 1/h(t) = 1/\lambda$ for $m = 1$.

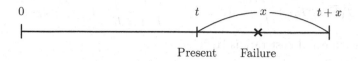

Fig. 5.1. Process of model with forward time

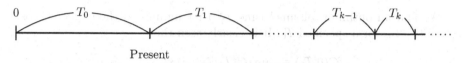

Fig. 5.2. Successive age replacement times

(3) If F is IFR (DFR), *i.e.*, $h(t)$ is increasing (decreasing), then $h(t) \leq (\geq)1/\alpha(t)$, and $\alpha(t)$ is decreasing (increasing), respectively, because

$$\frac{d\alpha(t)}{dt} = \alpha(t)\left[h(t) - \frac{1}{\alpha(t)}\right].$$

5.2 Age Replacement

Consider the age replacement policies when a unit has not failed in time T_0 $(0 \leq T_0 < \infty)$. Suppose that the unit is replaced at a planned time $T_0 + T_1$ $(0 < T_1 \leq \infty)$ or at failure, whichever occurs first, given that it is operating at time T_0. Let c_1 be the replacement cost for a failed unit and c_2 be the replacement cost at time $T_0 + T$ with $c_2 < c_1$.

A simple method of age replacement is to balance the costs for replacement after failure against that before failure, *i.e.*, $c_1 F(T_1|T_0) = c_2 \overline{F}(T_1|T_0)$. In this case,

$$F(T_1|T_0) = \frac{c_2}{c_1 + c_2}. \tag{5.6}$$

Using the entropy model from **(26)** in Sect. 9.3,

$$\frac{\log_2 F(T_1|T_0)}{\log_2 \overline{F}(T_1|T_0)} = \frac{c_1}{c_2}. \tag{5.7}$$

From this relation, we can easily calculate $F(T_1|T_0)$.

Next, the mean time from T_0 to replacement is

$$T_1 \overline{F}(T_1|T_0) + \int_0^{T_1} x \, dF(x|T_0) = \frac{1}{\overline{F}(T_0)} \int_0^{T_1} \overline{F}(T_0 + x) \, dx.$$

Thus, the expected cost rate is [1,2]

$$
\begin{aligned}
C(T_1|T_0) &= \frac{c_1 F(T_1|T_0) + c_2 \overline{F}(T_1|T_0)}{\int_0^{T_1} \overline{F}(x|T_0) \, dx} \\
&= \frac{c_1 \overline{F}(T_0) - (c_1 - c_2)\overline{F}(T_0 + T_1)}{\int_0^{T_1} \overline{F}(T_0 + x) \, dx}.
\end{aligned}
\tag{5.8}
$$

We find an optimum planned time T_1^* that minimizes the expected cost rate $C(T_1|T_0)$ for a specified T_0. It is clearly seen that

$$C(0|T_0) \equiv \lim_{T_1 \to 0} C(T_1|T_0) = \infty,$$

$$C(\infty|T_0) \equiv \lim_{T_1 \to \infty} C(T_1|T_0) = \frac{c_1}{\alpha(T_0)}. \tag{5.9}$$

Thus, there exists an optimum T_1^* ($0 < T_1^* \le \infty$) that minimizes $C(T_1|T_0)$. Differentiating $C(T_1|T_0)$ with respect to T_1 and setting it equal to zero,

$$h(T_0 + T_1) \int_0^{T_1} \overline{F}(T_0 + x) \, dx - [F(T_0 + T_1) - F(T_0)] = \frac{c_2 \overline{F}(T_0)}{c_1 - c_2}. \tag{5.10}$$

Letting $Q(T_1|T_0)$ be the left-hand side of (5.10),

$$Q(0|T_0) = 0,$$

$$Q(\infty|T_0) = h(\infty) \int_{T_0}^\infty \overline{F}(x) \, dx - \overline{F}(T_0),$$

where $h(\infty) \equiv \lim_{t \to \infty} h(t)$. In addition, if the failure rate $h(t)$ increases strictly, then $Q(T_1|T_0)$ also increases strictly with T_1 because for any $\Delta T > 0$,

$$h(T_0 + T_1 + \Delta T) \int_0^{T_1 + \Delta T} \overline{F}(T_0 + x) \, dx - F(T_0 + T_1 + \Delta T)$$

$$- h(T_0 + T_1) \int_0^{T_1} \overline{F}(T_0 + x) \, dx + F(T_0 + T_1)$$

$$\ge h(T_0 + T_1 + \Delta T) \int_0^{T_1 + \Delta T} \overline{F}(T_0 + x) \, dx$$

$$- h(T_0 + T_1 + \Delta T) \int_{T_1}^{T_1 + \Delta T} \overline{F}(T_0 + x) \, dx - h(T_0 + T_1) \int_0^{T_1} \overline{F}(T_0 + x) \, dx$$

$$= [h(T_0 + T_1 + \Delta T) - h(T_0 + T_1)] \int_0^{T_1} \overline{F}(T_0 + x) \, dx > 0.$$

Therefore, if $h(t)$ increases strictly and $h(\infty) > c_1/[(c_1 - c_2)\alpha(T_0)]$, then there exists a finite and unique T_1^* $(0 < T_1^* < \infty)$ that satisfies (5.10), and the resulting cost rate is

$$C(T_1^*|T_0) = (c_1 - c_2)h(T_0 + T_1^*). \tag{5.11}$$

Conversely, if $h(\infty) \leq c_1/[(c_1 - c_2)\alpha(T_0)]$, then $T_1^* = \infty$, i.e., the unit is replaced only at failure, and the expected cost rate is given in (5.9). When $T_0 = 0$, all results agree with those of the standard age replacement [1,2].

In general, it would be wasteful to replace an operating unit too early before failure because it might work longer and bring more profits. We introduce the modified replacement cost before failure: The replacement cost at time T when the unit will fail at time t $(t > T)$ is $c_2 + c_3(t - T)$. Then, the total expected cost until replacement is

$$C(T) = c_1 F(T) + \int_T^\infty [c_2 + c_3(t - T)]\, dF(t)$$

$$= c_1 F(T) + c_2 \overline{F}(T) + c_3 \int_T^\infty \overline{F}(t)\, dt. \tag{5.12}$$

It is clearly seen that

$$C(0) \equiv \lim_{T \to 0} C(T) = c_2 + c_3 \mu,$$
$$C(\infty) \equiv \lim_{T \to \infty} C(T) = c_1.$$

Differentiating $C(T)$ with respect to T and setting it equal to zero,

$$h(T) = \frac{c_3}{c_1 - c_2}. \tag{5.13}$$

Therefore, we have the optimum policy that minimizes $C(T)$ when $h(t)$ increases strictly:

(i) If $h(0) \geq c_3/(c_1 - c_2)$, then $T^* = 0$.
(ii) If $h(0) < c_3/(c_1 - c_2) < h(\infty)$, then there exists a finite and unique T^* $(0 < T^* < \infty)$ that satisfies (5.13).
(iii) If $h(\infty) \leq c_3/(c_1 - c_2)$, then $T^* = \infty$.

It is of interest in case (ii) that an operating unit should be replaced before failure when the failure rate attains a certain threshold level. In particular, when $F(t) = 1 - e^{-\lambda t^m}$ $(m > 1)$, i.e., $h(t) = \lambda m t^{m-1}$, an optimum replacement time is given by

$$T^* = \left[\frac{c_3}{\lambda m(c_1 - c_2)} \right]^{1/(m-1)}.$$

Note that T^* has the form similar to T_0 in [2, p. 98].

5.3 Reliability with Scheduling

The general definition of reliability is given by the probability that a unit continues to operate without failure during the interval $(t, t + x]$ when it is operating at time t, that is called *interval reliability* [1,2]. However, most units usually have to perform their functions for a job with working time.

Suppose that a job has a working time S such as operating time and processing time and should be achieved in time S by a unit. A job in the real world is done in random environment subject to many sources of uncertainty [110]. So that it would be reasonable to assume that S is a random variable and to define the reliability as the probability that the work of a job is accomplished successfully by a unit.

It is assumed that positive random variables S and X are the working time of a job and the failure time of the unit, respectively. Two random variables S and X are independent of each other and have the respective distributions $W(t)$ and $F(t)$ with finite means, *i.e.*, $W(t) \equiv \Pr\{S \le t\}$ and $F(t) \equiv \Pr\{X \le t\}$, where $\overline{\Phi}(t) \equiv 1 - \Phi(t)$ for any function $\Phi(t)$.

We define the reliability with working time as

$$R(W) \equiv \Pr\{S \le X\} = \int_0^\infty W(t) \, \mathrm{d}F(t) = \int_0^\infty \overline{F}(t) \, \mathrm{d}W(t), \qquad (5.14)$$

that represents the probability that the work of a job is accomplished by the unit without failure. The reliability $R(W)$ was defined as the expected gain with some weight function $W(t)$ [2]. If X and S are replaced with the strength and stress of some unit, respectively, this corresponds to the stress-strength model [111]. We have the following properties of $R(W)$ [2,107]:

(1) When $W(t)$ is the degenerate distribution placing unit mass at time T, $R(W) = \overline{F}(T)$ that is the usual reliability function. Furthermore, when $W(t)$ is a discrete distribution

$$W(t) = \begin{cases} 0 & \text{for } 0 \le t < T_1, \\ \sum_{i=1}^j p_i & \text{for } T_j \le t < T_{j+1} \quad (j = 1, 2, \ldots, N-1), \\ 1 & \text{for } t \ge T_N \end{cases}$$

$$R(W) = \sum_{j=1}^N p_j \overline{F}(T_j). \qquad (5.15)$$

(2) When $W(t) = F(t)$ for all $t \ge 0$, $R(W) = 1/2$.
(3) When $W(t) = 1 - \mathrm{e}^{-\omega t}$, $R(W) = 1 - F^*(\omega)$, where $\Phi^*(s)$ is the Laplace-Stieltjes transform of any function $\Phi(t)$, *i.e.*, $\Phi^*(s) \equiv \int_0^\infty \mathrm{e}^{-st} \, \mathrm{d}\Phi(t)$ for $s > 0$. Conversely, when $F(t) = 1 - \mathrm{e}^{-\lambda t}$, $R(W) = W^*(\lambda)$.
(4) When both S and X are normally distributed with mean μ_1 and μ_2 and variance σ_1^2 and σ_2^2, respectively, $R(W)$ is normally distributed with mean $\mu_2 - \mu_1$ and variance $\sigma_1^2 + \sigma_2^2$.

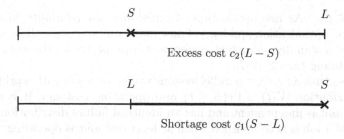

Fig. 5.3. Excess cost and shortage cost of scheduling

(5) When S is uniformly distributed during $[0, T]$, $R(W) = \int_0^T \overline{F}(t)\,dt/T$, that represents the interval availability for a finite interval $[0, T]$.

Some parts of a job need to be set up based on scheduling time [110]. If the work is not accomplished up to scheduling time, its time is prolonged, and this causes a great loss to scheduling. Conversely, if the work is accomplished too early before the scheduling time, this involves a waste of time or cost. The problem is how to determine the scheduling time of a job in advance.

It is assumed that a job has a working time S with a general distribution $W(t)$ with a finite mean $1/\omega$, and its scheduling time is L $(0 \le L < \infty)$ whose cost is $c_0(L)$. If the work is accomplished up to time L, *i.e.*, $L \ge S$, it needs the excess cost $c_2(L - S)$, and if it is not accomplished before time L and is done after L, *i.e.*, $L < S$, it needs the shortage cost $c_1(S - L)$ (Fig. 5.3). Then, the total expected cost until the work completion is

$$C(L) = \int_L^\infty c_1(t - L)\,dW(t) + \int_0^L c_2(L - t)\,dW(t) + c_0(L). \tag{5.16}$$

When $c_i(t) = c_i t$ and $c_i > 0$ $(i = 0, 1, 2)$, the total expected cost is

$$C(L) = c_1 \int_L^\infty \overline{W}(t)\,dt + c_2 \int_0^L W(t)\,dt + c_0 L. \tag{5.17}$$

We find an optimum time L^* that minimizes $C(L)$. It is clearly seen that there exists a finite L^* $(0 \le L^* < \infty)$ because $C(0) = c_1/\omega$ and $C(\infty) = \infty$. Differentiating $C(L)$ with respect to L and setting it equal to zero,

$$W(L) = \frac{c_1 - c_0}{c_1 + c_2}. \tag{5.18}$$

Therefore, we have the following optimum policy:

(i) If $c_1 > c_0$, then there exists a finite and unique L^* $(0 < L^* < \infty)$ that satisfies (5.18).

(ii) If $c_1 \le c_0$, then $L^* = 0$.

Example 5.1.   As one application of scheduling for reliability models, we consider the optimization problems of how many units for a parallel redundant system and a standby redundant system are appropriate for the work of a job with scheduling time S [107].

Suppose that an n-unit parallel system works for a job with working time S, its distribution $W(t) \equiv \Pr\{S \le t\}$ and operating cost $c_0 n$. It is assumed that each unit is independent and has an identical failure distribution $F(t)$. If the work of a job is accomplished when at least one unit is operating, it needs cost c_2, and if the work is accomplished after all n units have failed, it needs cost c_1 with $c_1 > c_2$. Then, the expected cost for an n-unit parallel system is

$$C_1(n) = c_2 + (c_1 - c_2) \int_0^\infty \overline{W}(t)\, d[F(t)]^n + c_0 n \qquad (n = 0, 1, 2, \dots). \quad (5.19)$$

There exists a finite number n^* $(0 \le n^* < \infty)$ that minimizes $C_1(n)$ because $C_1(0) = c_1$ and $C_1(\infty) = \infty$.

From the inequality $C_1(n+1) - C_1(n) \ge 0$,

$$\int_0^\infty [F(t)]^n \overline{F}(t)\, dW(t) \le \frac{c_0}{c_1 - c_2} \qquad (n = 0, 1, 2, \dots) \quad (5.20)$$

whose left-hand side decreases strictly to 0. Therefore, we have the following optimum policy:

(i) If $\int_0^\infty \overline{F}(t)\, dW(t) > c_0/(c_1 - c_2)$, then there exists a finite and unique minimum n^* $(1 \le n^* < \infty)$ that satisfies (5.20).
(ii) If $\int_0^\infty \overline{F}(t)\, dW(t) \le c_0/(c_1 - c_2)$, then $n^* = 0$, i.e., any parallel system should not be provided that might not reflect the actual situation of scheduling.

In particular, when $W(t) = 1 - e^{-\omega t}$ and $F(t) = 1 - e^{-\lambda t}$, (5.20) is

$$\sum_{j=0}^n \binom{n}{j}(-1)^j \frac{w}{w + (j+1)\lambda} \le \frac{c_0}{c_1 - c_2}.$$

If $w/(w + \lambda) > c_0/(c_1 - c_2)$, then there exists a positive n^* $(1 \le n^* < \infty)$. Table 5.1 presents the optimum number n^* of units for $c_0/(c_1 - c_2)$ and λ/w.

Next, consider a standby system with one operating unit and $n-1$ identical spare units, where each failed unit is replaced successively with one of the spare units. The system fails when all n units have failed. Suppose that $F^{(j)}(t)$ is the j-fold Stieltjes convolution of $F(t)$ with itself, and $F^*(s) \equiv \int_0^\infty e^{-st}dF(t)$ for $s > 0$ is the Laplace-Stieltjes transform of $F(t)$, i.e., $F^{(j)}(t) \equiv \int_0^t F^{(j-1)}(t - u)dF(u)$ $(j = 1, 2, \dots)$, $F^{(0)}(t) \equiv 1$ for $t \ge 0$, and $\int_0^\infty e^{-st}dF^{(j)}(t) = [F^*(s)]^j$ $(j = 0, 1, 2, \dots)$.

Table 5.1. Optimum numbers n^* of units for parallel and standby systems

$c_0/(c_1 - c_2)$	λ/w (parallel)			λ/w (standby)		
	1.0	2.0	5.0	1.0	2.0	5.0
0.5	0	0	0	0	0	0
0.3	1	1	0	1	1	0
0.1	2	2	1	3	3	3
0.05	3	4	2	4	5	7
0.01	9	12	11	6	9	16

Replacing $[F(t)]^n$ in (5.19) with $F^{(n)}(t)$ formally because the probability that all n units for an n-unit standby system have failed until time t is $F^{(n)}(t)$, the expected cost is

$$C_2(n) = c_2 + (c_1 - c_2) \int_0^\infty \overline{W}(t)\, dF^{(n)}(t) + c_0 n \quad (n = 0, 1, 2, \dots). \quad (5.21)$$

In particular, when $W(t) = 1 - e^{-wt}$,

$$C_2(n) = c_2 + (c_1 - c_2)[F^*(w)]^n + c_0 n \qquad (n = 0, 1, 2, \dots). \quad (5.22)$$

From the inequality $C_2(n+1) - C_2(n) \geq 0$,

$$[F^*(w)]^n [1 - F^*(w)] \leq \frac{c_0}{c_1 - c_2} \qquad (n = 0, 1, 2, \dots). \quad (5.23)$$

Therefore, we have the optimum policy:

(iii) If $F^*(w) < (c_1 - c_2 - c_0)/(c_1 - c_2)$, then there exists a finite and unique minimum n^* $(1 \leq n^* < \infty)$ that satisfies (5.23).

(iv) If $F^*(w) \geq (c_1 - c_2 - c_0)/(c_1 - c_2)$, then $n^* = 0$.

In addition, when $F(t) = 1 - e^{-\lambda t}$, (5.23) is

$$\frac{w\lambda^n}{(w + \lambda)^{n+1}} \leq \frac{c_0}{c_1 - c_2}.$$

If $w/(w+\lambda) > c_0/(c_1 - c_2)$, then a positive n^* $(1 \leq n < \infty)$ exists, that is the same as the result of a parallel system. Table 5.1 also presents the optimum number n^* that increases with $c_0/(c_1 - c_2)$. The optimum numbers n^* of a standby system are equal to or greater than those of a parallel system for large $c_0/(c_1 - c_2)$. Furthermore, the probability that the work is not accomplished by the system with n units is equal to or less than p is given by replacing $c_0/(c_1 - c_2)$ with p. ∎

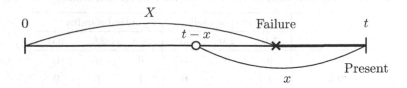

Fig. 5.4. Process of model with backward time

5.4 Backward Time

Suppose that a unit begins to operate at time 0, and X denotes its failure time with a distribution $F(t) \equiv \Pr\{X \le t\}$, a density function $f(t)$, and its mean $\mu \equiv E\{X\} = \int_0^\infty t f(t) \mathrm{d}t = \int_0^\infty \overline{F}(t) \mathrm{d}t$, where $\overline{F}(t) \equiv 1 - F(t)$. Then, the probability that a unit failed during $(t - x, t]$ $(0 \le x \le t)$, given that it is detected to have failed at time t, is

$$H(x|t) \equiv \Pr\{t - x \le X | X \le t\} = \frac{F(t) - F(t - x)}{F(t)} \le 1 \tag{5.24}$$

for $F(t) > 0$ (Fig. 5.4). In this case, we call $t - X$ *backward time* that is called *waiting time* [112]. Because

$$\overline{H}(x|t) \equiv 1 - H(x|t) = \frac{F(t - x)}{F(t)}, \tag{5.25}$$

its density function is

$$r(x|t) \equiv \frac{\mathrm{d}H(x|t)}{\mathrm{d}x} = \frac{f(t - x)}{F(t)}, \tag{5.26}$$

and its mean backward time, *i.e.*, the mean time from t to the failure time is

$$\beta(t) \equiv E\{t - X | X \le t\} = \int_0^t \overline{H}(x|t) \, \mathrm{d}x = \frac{1}{F(t)} \int_0^t F(x) \, \mathrm{d}x. \tag{5.27}$$

We summarize the properties of $H(x|t)$, $r(x|t)$, and $\beta(t)$ [112–114]:

(1) When $f(t)$ is continuous, $H(x|t)$ increases from $H(0|t) = 0$ to $H(t|t) = 1$, *i.e.*, $H(x|t)$ is a proper distribution for $0 \le x \le t$. Furthermore,

$$r(t) \equiv r(0|t) = \frac{f(t)}{F(t)}, \tag{5.28}$$

that is called *reversed failure (hazard) rate*, and $r(t)\mathrm{d}t$ represents the probability of failure in $(t - \mathrm{d}t, t]$, given that the unit is detected to have failed at time t. Let $H(t) \equiv \int_t^\infty r(u) du$ that is called *reversed cumulative failure (hazard) rate*. Because $F'(t)/F(t) = r(t)$ and $F(\infty) \equiv \lim_{t \to \infty} F(t) = 1$, clearly

$$F(t) = e^{-\int_t^\infty r(u) du}, \tag{5.29}$$

$$\overline{H}(x|t) = e^{-\int_{t-x}^t r(u) du}. \tag{5.30}$$

Moreover, because

$$r(t) = \frac{1}{F(t)} \lim_{x \to 0} \frac{F(t) - F(t-x)}{x},$$

it is easily noted that if $H(x|t)$ is decreasing (increasing) in t for $x > 0$, then $r(t)$ is decreasing (increasing). Conversely, if $r(t)$ is decreasing, then $r(x_1) \geq r(x_2)$ for $x_1 \leq x_2$, and hence,

$$\int_0^t r(x_1 - u) \, du \geq \int_0^t r(x_2 - u) \, du,$$

i.e.,

$$\exp\left[-\int_{x_2-t}^{x_2} r(u) \, du\right] \geq \exp\left[-\int_{x_1-t}^{x_1} r(u) \, du\right].$$

Thus, $\overline{H}(t|x_2) \geq \overline{H}(t|x_1)$, that implies that $H(x|t)$ decreases with t. Similarly, if $r(t)$ increases, then $H(x|t)$ also increases with t. From the above discussions, both $r(t)$ and $H(x|t)$ have the same monotonic properties for t. The important result was proved [114] that non-negative random variables cannot have increasing reversed failure rates from (5.29).

(2) If $r(t)$ is decreasing (increasing), then

$$\frac{F(t)}{\int_0^t F(x) \, dx} \geq (\leq) r(t). \tag{5.31}$$

Thus, when $r(t)$ decreases, $\beta(t)$ increases from $1/r(0)$ to ∞.

(3) When $F(t) = 1 - e^{-\lambda t}$,

$$H(x|t) = \frac{e^{\lambda x} - 1}{e^{\lambda t} - 1},$$

$$r(t) = \frac{\lambda}{e^{\lambda t} - 1}, \qquad H(t) = -\log(1 - e^{-\lambda t}),$$

$$\beta(t) = \frac{t}{1 - e^{-\lambda t}} - \frac{1}{\lambda}.$$

Thus, all of $H(x|t)$, $r(t)$, and $H(t)$ decrease with t from ∞ to 0, and $\beta(t)$ increases from 0 to ∞. Because $e^a \approx 1 + a$ for small a,

$$H(x|t) \approx \frac{x}{t}, \qquad r(t) \approx \frac{1}{t}$$

for $0 \leq x \leq t$, that is approximately distributed uniformly in $[0, t]$, and $\beta(t) \approx t/2$.

(4) When $F(t) = 1 - e^{-\lambda t^m}$ $(m > 0)$,

$$r(t) = \frac{\lambda m t^{m-1}}{e^{\lambda t^m} - 1},$$

$$\frac{dr(t)}{dt} = \frac{\lambda m t^{m-2} e^{\lambda t^m}}{(e^{\lambda t^m} - 1)^2}[(m-1)(1 - e^{-\lambda t^m}) - \lambda m t^m]$$

$$\leq \frac{\lambda m t^{m-2} e^{\lambda t^m}}{(e^{\lambda t^m} - 1)^2}[(m-1)\lambda t^m - \lambda m t^m] < 0.$$

Thus, the reversed failure rate $r(t)$ decreases strictly from ∞ to 0 for any $m > 0$.

The reversed failure rate was first defined [115], and some results on its ordering were obtained [116, 117]. The monotonic properties of the reversed failure rate were investigated [112–114]. However, there is no paper that has been related to the reversed failure rate with maintenance models.

5.4.1 Optimum Backward Times

Consider some problems of obtaining an optimum backward time [108]. Suppose that when a unit has failed at time t $(0 < t < \infty)$, and its failure time is unknown, we go back to time T $(0 \leq T \leq t)$ from time t to detect its failure, and call T a *planned backward time*. Then, the probability that the failure is detected in time T is p $(0 \leq p \leq 1)$, *i.e.*, the pth percentile point T_p of distribution $H(x|t)$ in (5.24) is given by

$$H(T_p|t) = \frac{F(t) - F(t - T_p)}{F(t)} = p. \qquad (5.32)$$

When $F(t) = 1 - e^{-\lambda t}$,

$$T_p = \frac{1}{\lambda} \log(1 - p + pe^{\lambda t}).$$

For example, when $t = 1/\lambda = 100$ and $p = 0.90$, $T_p = 93.47$.

In addition, we introduce the following costs (Fig. 5.5): Cost $c_1(x)$ is the excess cost suffered for the time x from a failure to the backward time, $c_2(x)$ is the shortage cost suffered for the time x from the backward time to a failure, and $c_0(T)$ is the cost required for the backward time T, where $c_0(0) \equiv 0$; this cost includes all costs resulting from the preparation and execution of the planned backward operation. Using the definition of $H(x|t)$, the total expected cost for the backward time T is

$$C(T|t) = \int_0^T c_1(T - x)\,dH(x|t) + \int_T^t c_2(x - T)\,dH(x|t) + c_0(T)$$

$$= \frac{1}{F(t)}\left[\int_{t-T}^t c_1(x - t + T)\,dF(x) + \int_0^{t-T} c_2(t - T - x)\,dF(x)\right] + c_0(T).$$

$$(5.33)$$

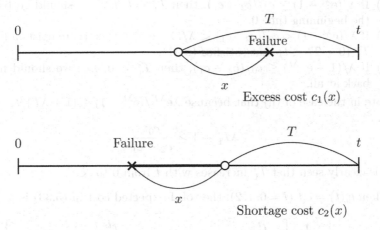

Fig. 5.5. Excess cost and shortage cost of backward time T

It has been well-known in a Poisson process that the occurrence of an event, given that there was an event in $[0, t]$, is uniformly distributed over $[0, t]$ [17, p. 71]. From this viewpoint when $H(x|t)$ is distributed uniformly in $[0, t]$, i.e., $H(x|t) = x/t$ for $0 \le x \le t$, the total expected cost is

$$C(T|t) = \frac{1}{t} \int_0^T c_1(x)\, dx + \frac{1}{t} \int_0^{t-T} c_2(x)\, dx + c_0(T). \tag{5.34}$$

We find an optimum time T^* that minimizes $C(T|t)$ for a given $t > 0$ in (5.33) in the following two cases:

(1) When $c_i(t) = c_i$ $(i = 1, 2)$ with $c_2 > c_1$ and $c_0(T) = c_0 T$, the total expected cost in (5.33) is

$$C_1(T|t) = c_1 + (c_2 - c_1)\frac{F(t - T)}{F(t)} + c_0 T. \tag{5.35}$$

Clearly,
$$C_1(0|t) = c_2, \qquad C_1(t|t) = c_1 + c_0 t.$$

Differentiating $C_1(T|t)$ in (5.35) with respect to T and setting it equal to zero,

$$r(T|t) = \frac{f(t - T)}{F(t)} = \frac{c_0}{c_2 - c_1}. \tag{5.36}$$

In the particular case of $F(t) = 1 - e^{-\lambda t}$, (5.36) is rewritten as

$$\frac{\lambda e^{\lambda T}}{e^{\lambda t} - 1} = \frac{c_0}{c_2 - c_1}. \tag{5.37}$$

Therefore, we have the following optimum backward time T_1^*:

(i) If $\lambda/(e^{\lambda t} - 1) \geq c_0/(c_2 - c_1)$, then $T_1^* = t$, i.e., we should go back to the beginning time 0.

(ii) If $\lambda/(e^{\lambda t} - 1) < c_0/(c_2 - c_1) < \lambda/(1 - e^{-\lambda t})$, then there exists a unique T_1^* $(0 < T_1^* < t)$ that satisfies (5.37).

(iii) If $\lambda/(1 - e^{-\lambda t}) \leq c_0/(c_2 - c_1)$, then $T_1^* = 0$, i.e., we should not go back at all.

Note in the case of (ii) that because $\lambda e^{\lambda T}/(e^{\lambda t} - 1) \leq (1 + \lambda T)/t$,

$$\lambda T_1^* + 1 \geq \frac{c_0 t}{c_2 - c_1}.$$

It is clearly seen that T_1^* increases with t from 0 to ∞.

(2) When $c_i(t) = c_i t$ $(i = 0, 1, 2)$, the total expected cost in (5.33) is

$$C_2(T|t) = \frac{1}{F(t)} \left[c_1 \int_{t-T}^{t} (x - t + T)\, dF(x) + c_2 \int_0^{t-T} (t - T - x)\, dF(x) \right] + c_0 T$$

$$= \frac{1}{F(t)} \left[-c_1 \int_{t-T}^{t} F(x)\, dx + c_2 \int_0^{t-T} F(x)\, dx \right] + (c_1 + c_0)T. \quad (5.38)$$

Clearly,

$$C_2(0|t) = \frac{c_2}{F(t)} \int_0^t F(x)\, dx = c_2 \beta(t),$$

$$C_2(t|t) = (c_1 + c_0)t - \frac{c_1}{F(t)} \int_0^t F(x)\, dx = c_0 t + c_1 [t - \beta(t)].$$

Differentiating $C_2(T|t)$ with respect to T and setting it equal to zero,

$$H(T|t) = \frac{F(t) - F(t - T)}{F(t)} = \frac{c_2 - c_0}{c_2 + c_1}. \quad (5.39)$$

Thus, we may obtain a $p[= (c_2 - c_0)/(c_2 + c_1)]$th percentile point of distribution $H(x|t)$ and have the following optimum backward time T_2^*:

(i) If $c_2 > c_0$, then there exists a unique T_2^* $(0 < T_2^* < t)$ that satisfies (5.39).

(ii) If $c_2 \leq c_0$, then $T_2^* = 0$, i.e., we should not go back at all.

In particular, when $F(t) = 1 - e^{-\lambda t}$ and $c_2 > c_0$, T_2^* is a unique solution of the equation

$$\frac{e^{\lambda T} - 1}{e^{\lambda t} - 1} = \frac{c_2 - c_0}{c_2 + c_1}. \quad (5.40)$$

Because $e^a \approx 1 + a$ for small a, T_2^* is given approximately by

$$\widetilde{T}_2 = \frac{c_2 - c_0}{c_2 + c_1} t, \quad (5.41)$$

that is equal to the optimum time that minimizes the expected cost $C_1(T|t)$ in (5.34) given by

$$C(T|t) = \frac{c_1 T^2 + c_2(t - T)^2}{2t} + c_0 T. \tag{5.42}$$

Further, because $(e^{\lambda t} - 1)/t \geq (e^{\lambda T} - 1)/T$ for $0 < T \leq t$, $T_2^* \geq \tilde{T}_2$, i.e., $T_2^* \geq [(c_2 - c_0)t]/(c_2 + c_1)$. If c_2 becomes larger, then T_2^* increases from 0 to t and also increases with t from 0 to ∞.

Example 5.2.   Table 5.2 presents the optimum times λT_2^* and approximate times $\lambda \tilde{T}_2$ for $\lambda t = 0.1$–2.0 and $c_2/c_1 = 1, 2, 5$, and 10 when $c_0/c_1 = 0.5$. For example, when $c_2/c_1 = 2$ and $\lambda t = 1$, i.e., a unit has failed at a mean failure time $1/\lambda$, we should go back 0.62 time to detect its failure. This indicates that T_2^* increases with t and c_2 and $T_2^* > \tilde{T}_2$. If c_2 becomes larger, then T_2^* approaches to time t, and the lower bound $\tilde{T}_2$ is a good approximation to T_2^* for small λt. ∎

Next, we consider the periodic inspection model [1, p. 201], where a unit is checked only at times jT $(j = 1, 2, \dots)$, and the inspection is not perfect, i.e., all failures cannot always be detected upon inspection, and undetected failures will occur at some later check [118]. The problem is how many checks we go back to detect its failure, given that it is detected at time KT $(K = 1, 2, \dots)$. This corresponds to a discrete optimization problem of the previous one.

Suppose that we go back N checks $(N = 0, 1, \dots, K)$ from KT when the failure is detected at time KT. The total expected cost for the backward number N, when costs c_i $(i = 1, 2, 3)$ are given by the function of inspection number, i.e., $c_i(j) = c_i j$ $(i = 0, 1, 2)$ in (2), is

$$C(N|K) = \frac{1}{F(KT)} \left\{ c_1 \sum_{j=K-N}^{K-1} \int_{jT}^{(j+1)T} [j - (K - N)] \, dF(t) \right.$$

$$\left. + c_2 \sum_{j=0}^{K-N} \int_{jT}^{(j+1)T} (K - N - j) \, dF(t) \right\} + c_0 N$$

$$= \frac{1}{F(KT)} \left\{ -c_1 \sum_{j=K-N+1}^{K} F(jT) + c_2 \sum_{j=1}^{K-N} F(jT) \right\} + (c_1 + c_0)N$$

$$(N = 0, 1, 2, \dots, K). \tag{5.43}$$

Clearly,

$$C(0|K) = \frac{c_2}{F(KT)} \sum_{j=1}^{K} F(jT),$$

$$C(K|K) = (c_1 + c_0)K - \frac{c_1}{F(KT)} \sum_{j=1}^{K} F(jT).$$

Table 5.2. Optimum time λT_2^* and approximate time $\lambda \widetilde{T}_2$ when $c_0/c_1 = 0.5$.

λt	c_2/c_1							
	1		2		5		10	
	λT_2^*	$\lambda \widetilde{T}_2$	λT_2^*	$\lambda \widetilde{T}_2$	λT_2^*	$\lambda \widetilde{T}_2$	λT_2^*	$\lambda \widetilde{T}_2$
0.1	0.026	0.025	0.051	0.050	0.076	0.075	0.087	0.086
0.2	0.054	0.050	0.105	0.100	0.154	0.150	0.175	0.173
0.5	0.150	0.125	0.281	0.250	0.396	0.375	0.445	0.432
1.0	0.357	0.250	0.620	0.500	0.828	0.750	0.910	0.864
1.5	0.626	0.375	1.001	0.750	1.284	1.125	1.388	1.295
2.0	0.954	0.500	1.434	1.000	1.756	1.500	1.875	1.727

From the inequality $C(N + 1|K) - C(N|K) \geq 0$,

$$H(NT|KT) = \frac{F(KT) - F(KT - NT)}{F(KT)} \geq \frac{c_2 - c_0}{c_2 + c_1} \qquad (N = 0, 1, 2, \ldots, K).$$
(5.44)

Thus, by the similar method for obtaining (i) and (ii), we have the optimum policy:

(i) If $c_2 > c_0$, then there exists a unique minimum number N^* ($N^* = 1, 2, \ldots, K$) that satisfies (5.44).

(ii) If $c_2 \leq c_0$, then $N^* = 0$.

5.4.2 Traceability

We apply backward time to the traceability problem used commonly in food products [119]. Suppose that a unit is detected to have failed at time T by some check (Fig. 5.6). We consider two cases: One is that operational behaviors of the unit are on record, and its failure time can be traced back easily and be known. In this case, when the unit has a failure distribution $F(t)$ with a finite mean μ, we introduce the following simple expected cost as one typical objective function:

$$C_1(T) = c_0 T + c_1 + c_2 \int_0^T (T - t) \, dF(t)$$

$$= c_0 T + c_1 + c_2 \int_0^T F(t) \, dt,$$
(5.45)

where c_0 = tracing cost per unit of time, c_1 = the cost for one check, and c_2 = loss cost per unit of time from a failure to its detection and its search.

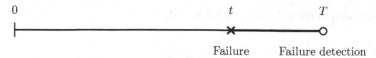

Failure   Failure detection

Fig. 5.6. Process of traceability

Thus, the expected cost rate until time T is given by

$$\widetilde{C}_1(T) \equiv \frac{C_1(T)}{T} = c_0 + \frac{c_1 + c_2 \int_0^T F(t)\, dt}{T}. \tag{5.46}$$

Because $\lim_{T \to 0} \widetilde{C}_1(T) = \infty$ and $\lim_{T \to \infty} \widetilde{C}_1(T) = c_0 + c_2$, there exists a positive T^* $(0 < T^* \le \infty)$ that minimizes $\widetilde{C}_1(T)$. Furthermore, differentiating $\widetilde{C}_1(T)$ with respect to T and setting it equal to zero,

$$\int_0^T t\, dF(t) = \frac{c_1}{c_2}. \tag{5.47}$$

Therefore, if $c_2 \mu > c_1$, then there exists a finite and unique T^* $(0 < T^* < \infty)$ that satisfies (5.47).

Second, the behaviors of the unit are not on record, so that it is difficult to trace back its failure time. In this case, the expected cost is given by

$$C_2(T) = c_1 + c_3 \int_0^T (T - t)\, dF(t), \tag{5.48}$$

where $c_3 = $ cost per unit of time for a failure and its search with $c_3 > c_2$. Comparing (5.45) and (5.48), if

$$\frac{1}{T} \int_0^T F(t)\, dt > \frac{c_0}{c_3 - c_2}, \tag{5.49}$$

then $C_2(T) > C_1(T)$, *i.e.*, we should always trace the behaviors of the unit because the left-hand side of (5.49) increases strictly with T from 0 to 1. Therefore, if $c_3 > c_2 + c_0$ and T_1 is a finite and unique solution of the equation

$$\frac{1}{T} \int_0^T F(t)\, dt = \frac{c_0}{c_3 - c_2}, \tag{5.50}$$

then $C_2(T) > C_1(T)$ for $T > T_1$, *i.e.*, we should always trace the unit.

Example 5.3.   Suppose that the failure time is exponential, *i.e.*, $F(t) = 1 - e^{-\lambda t}$. If $c_2/\lambda > c_1$, then from (5.47), there exists a finite and unique T^* $(0 < T^* < \infty)$ that satisfies

$$\frac{1}{\lambda} \left[1 - (1 + \lambda T) e^{-\lambda T} \right] = \frac{c_1}{c_2}.$$

Using the inequality $e^{-a} > 1 - a$ for $a > 0$,

$$T^* > \sqrt{\frac{c_1}{\lambda c_2}}.$$

Furthermore, from (5.50), T_1 is a finite and unique solution of the equation

$$1 - \frac{1 - e^{-\lambda T}}{\lambda T} = \frac{c_0}{c_3 - c_2}$$

for $c_3 > c_2 + c_0$. Thus, if $T > T_1$, then we should trace the unit. In this case,

$$T > T_1 > \frac{2c_0}{\lambda(c_3 - c_2)}.$$

Next, suppose that $F(t)$ is distributed uniformly in $[0, T]$, *i.e.*, $F(t) = t/T$ for $0 \le t \le T$. Then, from (5.49), if $c_3 > c_2 + 2c_0$, then we should trace the unit. ∎

It has been assumed until now that c_0 is the tracing cost. Viewed from another angle, c_0 may be assumed to be insurance for some accidents. Then, this corresponds to the stochastic problem of whether or not we should insure for such objects.

5.5 Checking Interval

Most units in standby and storage have to be checked at planned times to detect their failures. Such inspection models have assumed that any failure is known only through checking and summarized [1, 2]. When a failure is detected in the recovery technique for a database system, we execute the rollback operation to the latest checkpoint [120, 121] and reconstruct the consistency of the database that will be dealt with in Chap. 7. It has been assumed in such models that any failure is always detected immediately, however, there is a loss of time or cost with the lapsed time for the rollback operation between the detection of a failure and the latest checkpoint.

From the practical viewpoints of backup operation and database recovery, we consider the backup model that is one of the modified inspection policies: When the failure is detected, we execute the backup operation to the latest checking time (Fig. 5.7). The problem is to determine an optimum schedule T_k^* of checking intervals.

It is assumed that the failure time of a unit has a probability distribution $F(t)$ with a finite mean μ, where $\overline{F}(t) \equiv 1 - F(t)$. The checking times are placed at successive times T_k ($k = 1, 2, \ldots$), where $T_0 \equiv 0$. Let c_1 be the cost required for each check. In addition, when the failure is detected between T_k and T_{k+1}, we execute the backup operation to the latest checking time T_k that incurs a cost $c_2(x)$, where $c_2(0) \equiv 0$.

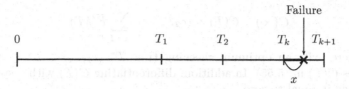

Fig. 5.7. Process of sequential checking intervals

The total expected cost until the backup operation is done to the latest checking time when a unit has failed is [2]

$$C(T_1, T_2, \ldots) = \sum_{k=0}^{\infty} \int_{T_k}^{T_{k+1}} [kc_1 + c_2(t - T_k)] \, dF(t)$$

$$= \sum_{k=1}^{\infty} [c_1 - c_2(T_k - T_{k-1})] \overline{F}(T_k) + \sum_{k=0}^{\infty} \int_0^{T_{k+1}-T_k} \overline{F}(t + T_k) \, dc_2(t). \quad (5.51)$$

If each check is done at periodic times kT $(k = 1, 2, \ldots)$, then

$$C(T) = [c_1 - c_2(T)] \sum_{k=1}^{\infty} \overline{F}(kT) + \sum_{k=0}^{\infty} \int_0^T \overline{F}(t + kT) \, dc_2(t). \quad (5.52)$$

When $c_2(t) = c_2 t$, the total expected cost is

$$C(T_1, T_2, \ldots) = \sum_{k=1}^{\infty} [c_1 - c_2(T_k - T_{k-1})] \overline{F}(T_k) + c_2 \mu. \quad (5.53)$$

Let $f(t)$ be a density function of $F(t)$. Then, differentiating $C(T_1, T_2, \ldots)$ with respect to T_k and setting it equal to zero,

$$\frac{F(T_{k+1}) - F(T_k)}{f(T_k)} = T_k - T_{k-1} - \frac{c_1}{c_2} \quad (k = 1, 2, \ldots). \quad (5.54)$$

Thus, we can compute the optimum checking times T_k^*, using Algorithm 1 of [2].

The total expected cost for periodic checking time is, from (5.52),

$$C(T) = (c_1 - c_2 T) \sum_{k=1}^{\infty} \overline{F}(kT) + c_2 \mu. \quad (5.55)$$

Clearly,

$$C(0) \equiv \lim_{T \to 0} C(T) = \infty, \qquad C(\infty) \equiv \lim_{T \to \infty} C(T) = c_2 \mu.$$

Hence,

$$C(\infty) - C(T) = (c_2 T - c_1) \sum_{k=1}^{\infty} \overline{F}(kT).$$

Thus, there exits an optimum checking time T^* ($c_1/c_2 < T^* \leq \infty$) that minimizes $C(T)$ in (5.55). In addition, differentiating $C(T)$ with respect to T and setting it equal to zero,

$$T - \frac{\sum_{k=1}^{\infty} \overline{F}(kT)}{\sum_{k=1}^{\infty} k f(kT)} = \frac{c_1}{c_2}. \tag{5.56}$$

In the case of $F(t) = 1 - e^{-\lambda t}$, (5.56) becomes simply

$$T - \frac{1 - e^{-\lambda T}}{\lambda} = \frac{c_1}{c_2}. \tag{5.57}$$

It can be easily seen that the left-hand side of (5.57) increases strictly from 0 to ∞, and hence, there exists a finite and unique T^* that satisfies (5.57). Using $e^a \approx 1 + a + a^2/2$ for small a, the optimum checking time is given approximately by

$$\widetilde{T} = \sqrt{\frac{2c_1}{\lambda c_2}}, \tag{5.58}$$

and $T^* > \widetilde{T}$.

Example 5.4. We compute the optimum checking times numerically when the failure time has a Weibull distribution. Table 5.3 presents the optimum schedule $\{T_k^*\}$ that satisfies (5.54), and $\delta_k \equiv T_{k+1}^* - T_k^*$ for $c_1/c_2 = 10, 20,$ and 30 when $F(t) = 1 - \exp[-(t/500)^2]$. It is shown that δ_k decreases with k.

Next, suppose that the failure time is uniformly distributed in $[0, S]$ ($0 < S < \infty$), i.e., $F(t) = t/S$ for $0 \leq t \leq S$, and 0 for $t > S$. Then, (5.54) becomes

$$T_{k+1} - T_k = T_k - T_{k-1} - \frac{c_1}{c_2}, \tag{5.59}$$

that is equal to that of [2, p. 113]: Solving for T_k,

$$T_k = k T_1 - \frac{k(k-1)}{2} \frac{c_1}{c_2}.$$

Setting $T_N = S$,

$$T_k = \frac{kS}{N} + k(N - k) \frac{c_1}{2c_2} \qquad (k = 0, 1, 2, \ldots, N).$$

From $T_{k+1} - T_k > 0$,

$$\frac{S}{N} + \frac{c_1}{2c_2}(N - 2k - 1) > 0.$$

When $k = N - 1$,

Table 5.3. Optimum checking times T_k^* and $\delta_k = T_{k+1}^* - T_k^*$ when $F(t) = 1 - \exp[-(t/500)^2]$

| | c_1/c_2 | | | | | |
| | 10 | | 20 | | 30 | |
k	T_k^*	δ_k	T_k^*	δ_k	T_k^*	δ_k
1	145.04	107.24	183.23	136.06	211.09	157.44
2	252.28	91.62	319.29	116.43	368.53	135.06
3	343.90	82.54	435.72	104.90	503.59	121.86
4	426.44	76.49	540.92	97.09	625.45	112.86
5	502.93	72.20	637.71	91.40	738.29	106.13
6	575.13	69.12	729.11	87.15	844.42	100.88
7	644.25	67.03	816.26	84.06	945.30	96.61
8	711.28	65.90	900.32	82.17	1041.91	93.04
9	777.18	65.89	982.49	81.92	1134.95	89.99
10	843.07		1064.41		1224.94	

$$\frac{N(N-1)}{2} < \frac{c_2}{c_1}, \quad i.e., \quad \frac{N(N+1)}{2} > \frac{c_2}{c_1},$$

that corresponds to the type of (3.5).

For example, when $S = 100$ and $c_1/c_2 = 10$, the checking number is $N^* = 4$ and $T_k^* = \{40, 70, 90, 100\}$. It would be trivial in the case of a uniform distribution that the optimum schedule is equal to the standard inspection model. ∎

5.6 Inspection for a Scale

Suppose that there is a process in which we weigh some product by a scale in the final stage of manufacturing to obtain its exact weight [105]. However, the scale occasionally becomes uncalibrated and produces inaccurate weights for individual products. To prevent such incorrect weights, the scale is checked every day. If the scale is detected to be uncalibrated at the inspection, then it is adjusted, and we reweigh some volume of products. Two modified models, where inspection activities involve adjustment operations and are executed only for detecting scale inaccuracy, were proposed [122, 123].

When we have many products to weigh every day, we can regard the volume of products to be weighed as continuous. Let t (> 0) denote the total volume of products to weigh every day. For example, when we weigh chemical products by a scale, we may denote t as the total expected chemical products

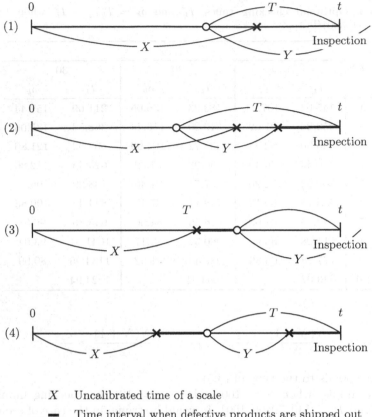

X   Uncalibrated time of a scale

—   Time interval when defective products are shipped out

Fig. 5.8. Four cases of a reweighing process for a scale

per day. When a scale is checked at only one time in the evening and is detected to be uncalibrated, some volume T of products are reweighed by the adjusted scale.

Let a non-negative random variable X be the time at which the scale becomes uncalibrated measured by the volume of weighed products. Hence, if $X > t$, then the scale is correct at the inspection, and all products are shipped out simultaneously. Conversely, if $X \leq t$, then the scale becomes uncalibrated. In this case, the scale is adjusted, and a volume T $(0 \leq T \leq t)$ of products is reweighed by this scale. In addition, let Y denote the time when the scale becomes inaccurate again, measured by the volume of reweighed products. If $Y < T$, then the scale becomes inaccurate again, and some defective products are shipped out.

Let U denote the volume of defective products to be shipped out. Then, we consider the following four cases (Fig. 5.8):

(i) $U = 0$ for $X > t$, or $t - T < X \le t$ and $Y > T$ in case (1).
(ii) $U = T - Y$ for $t - T < X \le t$ and $Y \le T$ in case (2).
(iii) $U = t - T - X$ for $X \le t - T$ and $Y > T$ in case (3).
(iv) $U = t - X - Y$ for $X \le t - T$ and $Y \le T$ in case (4).

It is assumed that X and Y are independent and identically distributed, and that both have an identical distribution $F(x)$. Then, from [105],

$$E\{U\} = F(t) \int_0^T (T - x)\, dF(x) + \int_0^{t-T} (t - T - x)\, dF(x)$$

$$= F(t) \int_0^T F(x)\, dx + \int_0^{t-T} F(x)\, dx. \qquad (5.60)$$

Next, let c_1 denote the cost incurred for shipping out a unit volume of defective products, and c_2 denote the cost for reweighing a unit volume of products. Then, the total expected cost during $(0, t]$, including the time for reweighing when the scale is uncalibrated, is

$$C(T|t) \equiv c_1 E\{U\} + c_2 T F(t)$$

$$= c_1 \left[F(t) \int_0^T F(x)\, dx + \int_0^{t-T} F(x)\, dx \right] + c_2 T F(t). \qquad (5.61)$$

Evidently,

$$C(0|t) = c_1 \int_0^t F(x)\, dx, \qquad C(t|t) = F(t) \left[c_1 \int_0^t F(x)\, dx + c_2 t \right].$$

Differentiating $C(T|t)$ with respect to T and setting it equal to zero,

$$\frac{F(t) - F(t - T)}{F(t)} + F(T) = \frac{c_1 - c_2}{c_1}, \qquad (5.62)$$

whose left-hand side increases strictly with T $(0 \le T \le t)$ from 0 to $1 + F(t)$. Therefore, we have the following optimum policy:

(i) If $c_1 > c_2$, then there exists a unique T^* $(0 < T^* < t)$ that satisfies (5.62).
(ii) If $c_1 \le c_2$, then $T^* = 0$, i.e., we should not reweigh any products.

Example 5.5.   Suppose that the failure time has a gamma distribution of order k, i.e., $\overline{F}(t) = \sum_{j=k}^{\infty} [(\lambda t)^j / j!] e^{-\lambda t}$. Table 5.4 presents the optimum T^* and expected cost $C(T^*|1)$ for $k = 1$, 2 and c_1/c_2 when $t = 1$ and $\lambda = 0.182$ for $k = 1$ and $\lambda = 0.73$ for $k = 2$, i.e., $F(t) \approx 1/6$ for both cases. It is observed that T^* increases from 0 to 0.865 for $k = 1$ and from 0 to 0.732 for $k = 2$ as c_1/c_2 increases from 0 to ∞. The optimum T^* for $k = 2$ are less than those for $k = 1$ and $C(T^*|1)/c_2$ for $k = 2$ is higher than those for $k = 1$ because the variance of a gamma distribution for $k = 2$ is two times that for $k = 1$. ∎

Table 5.4. Optimum T^* and expected cost $C(T^*|1)/c_2$

c_1/c_2	$k = 1$		$k = 2$	
	T^*	$C(T^*|1)/c_2$	T^*	$C(T^*|1)/c_2$
1	0.000	0.086	0.000	0.291
2	0.445	0.134	0.329	0.490
3	0.587	0.158	0.447	0.629
4	0.658	0.176	0.510	0.753
5	0.700	0.192	0.549	0.871
10	0.783	0.261	0.635	1.436
50	0.849	0.763	0.711	5.862
100	0.857	1.384	0.721	11.385
∞	0.865	∞	0.732	∞

6

Optimum Retrial Number of Reliability Models

We have often experienced in daily life making some trials of a system that result in either success or failure. If such a result becomes successful or fails, then it is said that the trial succeeds or fails, respectively. When the trial fails, we sometimes do the retrial, retry, reset, or restart. Such retrials are usually continued until the trial succeeds or stops at a limited number. It is well-known that the trial is called a *Bernoulli trial* and has a geometric distribution when each trial is independent and the probabilities of success or failure are constant, irrespective of its number [17]. However, the probability that the retrial succeeds would decrease generally with its number and time in actual situations, that is, the retrial process would have the DFR (Decreasing Failure Rate) property [1, p. 6], so that it might be wise from the viewpoints of economics and reliability that we stop the retrial when its total number exceeds a threshold level. In this case, it would be reasonable to investigate the origin of failure and also inspect and maintain the system by using reliability techniques, and to do the trial from the beginning.

We first introduce the standard stochastic retrial models in Sect. 6.1: We do the trial of some event for a system, and when it fails, we repeat consecutively the same N retrials including the first until it succeeds. When all N retrials have failed, we inspect and maintain the system, or switch it to another, and start the same trial from the beginning. We repeat such processes until the trial succeeds. It is assumed that the probability of the retrial succeeding at the jth number is q_j $(j = 1, 2, \ldots, N)$ that decreases with j. The mean time to success is obtained, and an optimum number N^* that minimizes it is discussed analytically. Next, we apply this model to the error detection scheme with checkpoints treated in Chap. 7 in detail. Last, it is shown that this forms a Markov renewal process, and the policy maximizing the availability is the same as minimizing the mean time to success theoretically.

It is important to estimate failure probabilities of trials. It is assumed in Sect. 6.2 that the probability that the trial fails at the jth number is constant p. The conjugated prior distribution of p is estimated to have a beta distribution from the past data. Then, an optimum number N^* that minimizes the mean time to success is discussed, and a numerical example is presented [124].

The most important problem in a communication system is how to transmit the data accurately and rapidly to a receiver. However, errors in data transmission are unavoidable because of disturbing factors such as disconnections, noises, or distortions in a communication line [125]. Error-control procedures are indispensable to transmit high quality data. The simplest scheme among error-control strategies is an automatic-repeat-request (ARQ) scheme in which a receiver requires the retransmission of the same data when errors have been detected [126, 127]. The ARQ strategy is widely employed in point-to-point data transmission because its error control is easy and simple. Retrial queues are stochastic models in which arrival jobs that find all servers busy repeat their attempts to get access to one of servers and describe the operation of telecommunication network [128].

We consider three ARQ models of data transmission with intermittent faults [129] as one application of the retrial model. The mean times to success of data transmission for the three models are obtained by using techniques of Markov renewal processes similar to Sect. 6.1 [33]. Optimum numbers N^* to minimize the mean times to success are discussed analytically, and some useful numerical examples are presented. Modified ARQ schemes of error-control procedures and their protocols were surveyed extensively [33].

6.1 Retrial Models

Consider the following three retrial models where we do a certain trial of some event:

(1) Standard Model

When the trial has failed, we repeat consecutively the same N retrials including the first one until it succeeds (Fig. 6.1). The probability that the retrial succeeds at the jth number is denoted by q_j that decreases strictly with j ($j = 1, 2, \ldots, N$), where $p_j \equiv 1 - q_j$. The time required for one trial is a mean time T_1. We call this cycle from the beginning of the trial to its success or all failures of N trials process 1. When all N retrials have failed, we do the same trial with the same success probability q_j in process 2 from the beginning after a mean time T_2. We repeat such processes until the trial succeeds.

Let $P(j) \equiv p_1 p_2 \ldots p_j$ ($j = 1, 2, \ldots, N$) and $P(0) \equiv 1$ be the probability that all j retrials have failed. The expected number of retrials until it succeeds in one process is

$$\sum_{j=1}^{N} j P(j-1) q_j = \sum_{j=0}^{N-1} P(j) - N P(N).$$

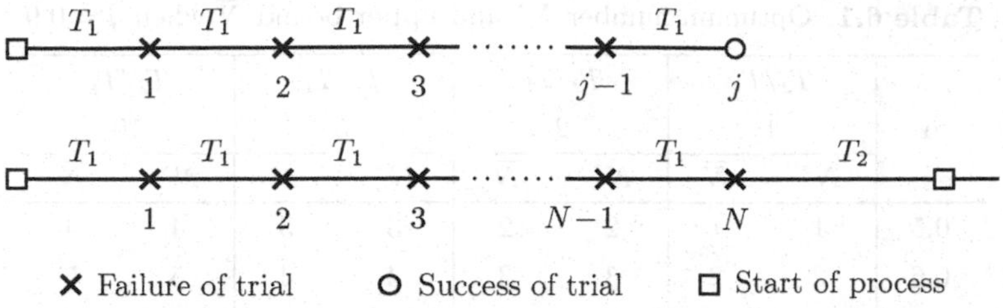

Fig. 6.1. Process of trials

The expected number of trials until success is given by a renewal equation:

$$M_1(N) = \sum_{j=0}^{N-1} P(j) - NP(N) + P(N)[N + M_1(N)],$$

i.e.,

$$M_1(N) = \frac{\sum_{j=0}^{N-1} P(j)}{1 - P(N)}. \tag{6.1}$$

Similarly, the mean time to success is given by a renewal equation:

$$\ell_1(N) = T_1 \left[\sum_{j=0}^{N-1} P(j) - NP(N) \right] + P(N)[NT_1 + T_2 + \ell_1(N)]. \tag{6.2}$$

Solving (6.2) for $\ell_1(N)$, the mean time to success is

$$\ell_1(N) = \frac{T_1 \sum_{j=0}^{N-1} P(j) + T_2 P(N)}{1 - P(N)} \qquad (N = 1, 2, \dots). \tag{6.3}$$

The mean time to success for most retrial models is represented by types of equations similar to (6.3).

We find an optimum number N^* that minimizes $\ell_1(N)$. From the inequality $\ell_1(N+1) - \ell_1(N) \geq 0$,

$$\frac{1 - P(N)}{q_{N+1}} - \sum_{j=0}^{N-1} P(j) \geq \frac{T_2}{T_1} \qquad (N = 1, 2, \dots). \tag{6.4}$$

Letting $L(N)$ be the left-hand side of (6.4),

$$L(N+1) - L(N) = [1 - P(N+1)] \left(\frac{1}{q_{N+2}} - \frac{1}{q_{N+1}} \right) > 0.$$

Thus, $L(N)$ increases strictly with N, and hence, if $L(\infty) \equiv \lim_{N \to \infty} L(N) > T_2/T_1$, then there exists a finite and unique minimum $N^*(1 \leq N^* < \infty)$ that satisfies (6.4).

Table 6.1. Optimum number N^* and upper bound $\overline{N}$ when $q = 0.9$

α	T_2/T_1 1		T_2/T_1 2		T_2/T_1 5		T_2/T_1 10	
	N^*	$\overline{N}$	N^*	$\overline{N}$	N^*	$\overline{N}$	N^*	$\overline{N}$
0.5	1	1	2	2	3	3	4	4
0.6	2	2	3	3	4	4	5	5
0.7	2	2	3	4	5	6	7	7
0.8	3	4	5	5	8	9	11	11
0.9	7	7	10	11	17	18	22	23

Furthermore, from the assumption that q_j decreases with j,

$$\frac{1 - P(N)}{q_{N+1}} - \sum_{j=0}^{N-1} P(j) \geq \frac{q_1}{q_{N+1}} - 1.$$

Thus, if there exists a unique minimum $\overline{N}$ such that $q_{N+1} \leq q_1 T_1/(T_1 + T_2)$, then $N^* \leq \overline{N}$. In addition, if $q_\infty \equiv \lim_{N \to \infty} q_N < q_1 T_1/(T_1 + T_2)$, then a finite N^* exists uniquely, and

$$\frac{T_1}{q_{N^*}} < \ell_1(N^*) + T_2 \leq \frac{T_1}{q_{N^*+1}}. \tag{6.5}$$

Example 6.1.   Suppose that $q_j = \alpha^{j-1} q$ $(0 < \alpha < 1, 0 < q < 1)$, *i.e.*, the probability of successful retrials decreases with its number at the rate of $100\alpha\%$. Then, (6.4) is rewritten as

$$\frac{1 - \prod_{j=1}^{N}(1 - \alpha^{j-1} q)}{\alpha^N q} - \sum_{j=0}^{N-1} \left[\prod_{i=1}^{j}(1 - \alpha^{i-1} q) \right] \geq \frac{T_2}{T_1} \quad (N = 1, 2, \dots), \tag{6.6}$$

where $\prod_{i=1}^{0} \equiv 1$. Because $\lim_{j \to \infty} q_j = 0$, there exists always a finite and unique minimum N^* $(1 \leq N^* < \infty)$ that satisfies (6.6). Clearly, if $\alpha \leq T_1/(T_1 + T_2)$, then $N^* = 1$. The upper bound $\overline{N}$ is given by a unique minimum such that $\alpha^N \leq T_1/(T_1 + T_2)$.

Table 6.1 presents the optimum number N^* and their upper bound $\overline{N}$ for α and T_2/T_1 when $q = 0.9$. The optimum N^* increases with both α and T_2/T_1 to ∞ as $\alpha \to 1$ or $T_2/T_1 \to \infty$. The upper bound $\overline{N}$ gives good approximations to N^* and would be used in actual fields as their rough estimations. ∎

(2) Checkpoint Model

We will take up the recovery process of error detection discussed in Chap. 7 [58]: Checkpoints are placed previously at periodic times kT $(k = 0, 1, 2, \dots)$

for a specified time $T > 0$. If some errors occur in an interval $((k-1)T, kT]$, we go back to the previous checkpoint time $(k-1)T$ and reexecute the process. It is assumed that error rates depend only on the total reexecution time in this interval including the first execution, irrespective of the numbers of errors and checkpoints, that is, when errors occur according to a general distribution $F(t)$, the probability that some errors occur at the jth reexecution in any checkpoint intervals is $[F(jT) - F((j-1)T)]/\overline{F}((j-1)T)$, where $\overline{F}(t) \equiv 1 - F(t)$. If no error occurs at some number of reexecutions, we say the process of this interval succeeds. In this case, the process can execute newly from the next checkpoint interval, and errors occur according to the same distribution $F(t)$.

We pay attention to only one of any checkpoint intervals. Suppose that the overhead time required for one reexecution due to errors is T_1. If errors occur in all N reexecutions, then we repeat the same process from the beginning after the overhead time T_2. We repeat such processes until the process succeeds. Then, the probability that some errors occur for all j reexecutions is

$$P(j) \equiv \prod_{i=1}^{j} \frac{F(iT) - F((i-1)T)}{\overline{F}((i-1)T)} \qquad (j = 1, 2, \ldots, N), \qquad (6.7)$$

and $P(0) \equiv 1$. Thus, substituting $P(j)$ in (6.3), we can obtain the mean time until the process succeeds in this checkpoint interval. In this case, the probability that no error occurs at the jth reexecution is

$$q_j = \frac{\overline{F}(jT)}{\overline{F}((j-1)T)} \qquad (j = 1, 2, \ldots).$$

Therefore, when the failure rate $[F(t+x) - F(t)]/\overline{F}(t)$ increases strictly with t for $x > 0$ and $\overline{F}(t) > 0$ [1, p. 6], q_j decreases strictly with j, and there exists a finite and unique minimum N^* that satisfies (6.4) if it exists. For example, when $\overline{F}(t) = \exp[-(\lambda t)^m]$ for $m > 1$, a finite N^* exists uniquely.

(3) Markov Renewal Process

We analyze the retrial model by using the techniques of Markov renewal processes [1, p. 28, 130, 131] (Fig. 6.2): A certain trial of some event happens according to a general distribution $F_0(t)$ with a mean T_0. We do the retrial consecutively until its success that needs the random time according to a general distribution $R(t)$ with a mean T_1. When all N retrials have failed, we do the same trial from the beginning after the random time according to a general distribution $G(t)$ with a mean T_2. It is assumed that the probability that the retrial succeeds at the jth number is q_j $(j = 1, 2, \ldots, N)$, the same as that of the previous model, where $p_j \equiv 1 - q_j$.

Under the above assumptions, we define the following states of the retrial process:

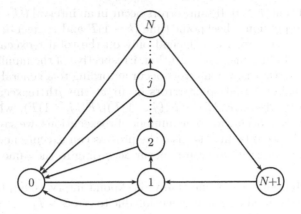

Fig. 6.2. State-transition diagram of a retrial model

State 0: An initial process begins.

State j: The jth number of retrials including the first one begins ($j = 1, 2, \ldots, N$).

State $N + 1$: The Nth number of retrials fails and the trial process begins newly after a mean time T_2.

The states defined above are regeneration points and form a Markov renewal process [130]. Let $Q_{ij}(t)$ ($i, j = 0, 1, 2, \ldots, N+1$) be one-step transition probabilities of a Markov renewal process. Then, mass functions $Q_{ij}(t)$ from State i to State j in an amount of time less than or equal to time t are

$$
\begin{aligned}
Q_{01}(t) &= F_0(t), \\
Q_{j0}(t) &= q_j R(t) && (j = 1, 2, \ldots, N), \\
Q_{jj+1}(t) &= p_j R(t) && (j = 1, 2, \ldots, N), \\
Q_{N+11}(t) &= G(t).
\end{aligned}
\tag{6.8}
$$

Let $H_{ij}(t)$ ($i, j = 0, 1$) be the first-passage time distribution from State i to State j. Then, from Fig. 6.2,

$$
H_{00}(t) = Q_{01}(t) * H_{10}(t),
\tag{6.9}
$$

$$
H_{10}(t) = \sum_{j=0}^{N-1} \left[\prod_{i=1}^{j} Q_{ii+1}(t) \right] * Q_{j+10}(t) + \left[\prod_{j=1}^{N} Q_{jj+1}(t) \right] * Q_{N+11}(t) * H_{10}(t),
\tag{6.10}
$$

where the asterisk denotes the Stieltjes convolution, i.e., $a(t) * b(t) \equiv \int_0^t b(t - u) \, da(u)$, $\prod_{i=1}^{j} Q_{ii+1}(t) \equiv Q_{12}(t) * \cdots * Q_{jj+1}(t)$ ($j = 1, 2, \ldots$), $\prod_{i=1}^{0} \equiv 1$, and $\Phi^*(s) \equiv \int_0^\infty e^{-st} d\Phi(t)$ for any function $\Phi(t)$. Taking the Laplace-Stieltjes transforms of (6.9) and (6.10) and substituting (6.8) in them,

$$H_{00}^*(s) \equiv \int_0^\infty e^{-st} \, dH_{00}(t) = \frac{F_0^*(s) \sum_{j=1}^N P(j-1) q_j [R^*(s)]^j}{1 - G^*(s) P(N) [R^*(s)]^N}. \qquad (6.11)$$

Thus, the mean recurrence time to State 0 is

$$\ell_{00}(N) \equiv \lim_{s \to 0} \frac{1 - H_{00}^*(s)}{s}$$

$$= T_0 + \frac{T_1 \sum_{j=0}^{N-1} P(j) + T_2 P(N)}{1 - P(N)} \qquad (N = 1, 2, \dots). \qquad (6.12)$$

Therefore, the optimum policy that minimizes $\ell_{00}(N)$ corresponds to the policy that minimizes $\ell_1(N)$ in (6.3).

Furthermore, we define the availability as the ratio of $\ell_1(N)/\ell_{00}(N)$, *i.e.*,

$$A(N) \equiv \frac{\ell_1(N)}{\ell_{00}(N)} = \frac{T_1 \sum_{j=0}^{N-1} P(j) + T_2 P(N)}{T_0[1 - P(N)] + T_1 \sum_{j=0}^{N-1} P(j) + T_2 P(N)}. \qquad (6.13)$$

It is naturally seen that the optimum policy that maximizes $A(N)$ also corresponds to that of minimizing $\ell_1(N)$ and $\ell_{00}(N)$.

6.2 Bayesian Estimation of Failure Probability

Suppose that the probability that the trial fails at j times ($j = 1, 2, \dots$) consecutively is $P(j) = p^j$ ($j = 1, 2, \dots$). Usually, we do not know the true value of p, however, we can estimate an approximate value of p from the past experiment. This section aims at the positive use of such *a priori* knowledge that is represented in a probability distribution based on Bayesian theory [124]. This is called the *priori distribution*. The conjugate prior distribution is often adopted from the viewpoint of simplicity of calculations for the posterior distribution when the data are obtained [132].

When the trial is done successively under the assumption of a constant failure probability p, the trial process forms basically a Bernoulli process. It is well-known that the conjugate prior distribution of p in such a case follows the beta distribution that is given by

$$f(p) = \frac{p^{\alpha-1}(1-p)^{\beta-1}}{B(\alpha, \beta)} \qquad \text{for } 0 \leq p \leq 1 \text{ and } \alpha, \beta > 0, \qquad (6.14)$$

where $B(\alpha, \beta) \equiv \Gamma(\alpha)\Gamma(\beta)/\Gamma(\alpha+\beta)$ and $\Gamma(x) \equiv \int_0^\infty e^{-t} t^{x-1} dt$ for $x > 0$.

To represent the prior knowledge of p, using the beta distribution in (6.14), it is necessary to set parameters α and β. There are various methods for setting two values of parameters. The simplest method is that the mean and variance of the beta distribution are, respectively,

$$E\{p\} = \frac{\alpha}{\alpha + \beta}, \qquad V\{p\} = \frac{\alpha\beta}{(\alpha + \beta)^2(\alpha + \beta + 1)}.$$

By estimating the prior mean and variance and solving them for α and β, two parameters can be determined.

The probability $\widehat{P}(j)$ that the trials fail for all j consecutive times can be estimated as, using the beta distribution,

$$\widehat{P}(j) \equiv \int_0^1 p^j f(p)\, dp = \frac{B(\alpha + j, \beta)}{B(\alpha, \beta)}$$
$$= \frac{\Gamma(\alpha + j)\Gamma(\alpha + \beta)}{\Gamma(\alpha + \beta + j)\Gamma(\alpha)} \qquad (j = 0, 1, 2, \dots). \tag{6.15}$$

Thus, the probability that the trial succeeds for the first time at the $(j+1)$th time is

$$\widehat{P}(j) - \widehat{P}(j+1) = \int_0^1 p^j(1-p)f(p)\, dp$$
$$= \frac{\beta\Gamma(\alpha + j)\Gamma(\alpha + \beta)}{\Gamma(\alpha + \beta + j + 1)\Gamma(\alpha)}. \tag{6.16}$$

Therefore, substituting (6.15) in (6.3), we can obtain the mean time $\ell_1(N)$ to success. In this case,

$$\widehat{q}_{N+1} = \frac{\widehat{P}(N) - \widehat{P}(N+1)}{\widehat{P}(N)} = \frac{\beta}{\alpha + \beta + N},$$

that decreases strictly with N to 0. Thus, from (6.4), there exists a finite and unique minimum N^* that satisfies

$$\frac{\alpha + \beta + N}{\beta}\left[1 - \frac{\Gamma(\alpha + \beta)}{\Gamma(\alpha)}\frac{\Gamma(\alpha + N)}{\Gamma(\alpha + \beta + N)}\right] - \frac{\Gamma(\alpha + \beta)}{\Gamma(\alpha)}\sum_{j=0}^{N-1}\frac{\Gamma(\alpha + j)}{\Gamma(\alpha + \beta + j)}$$
$$\geq \frac{T_2}{T_1} \qquad (N = 1, 2, \dots). \tag{6.17}$$

It is clearly seen that if $1/(\alpha + \beta) \geq T_2/T_1$, then $N^* = 1$.

Example 6.2. Suppose that two parameters of the beta distribution that is the conjugate prior distribution are $(\alpha, \beta) = (1, 9)$ and $(2, 18)$ [124]. When $(\alpha, \beta) = (1, 9)$, the mean and variance of p are

$$E\{p\} = \frac{1}{10}, \qquad V\{p\} = \frac{9}{1100},$$

Table 6.2. Optimum number N^* and mean time $\ell_1(N^*)/T_1$ when $(\alpha, \beta) = (1, 9)$ and $(\alpha, \beta) = (2, 18)$ [124]

T_2/T_1	$(\alpha, \beta) = (1, 9)$		$(\alpha, \beta) = (2, 18)$	
	N^*	$\ell_1(N^*)/T_1$	N^*	$\ell_1(N^*)/T_1$
0.1	1	1.122222	2	1.117391
0.2	2	1.124074	4	1.117627
0.3	3	1.124658	6	1.117645
0.4	4	1.124860	8	1.117647
0.5	5	1.124938	10	1.117647
0.6	6	1.124970	11	1.117647
0.7	7	1.124985	13	1.117647
0.8	8	1.124992	15	1.117647
0.9	9	1.124995	17	1.117647
1.0	10	1.124997	19	1.117647

and when $(\alpha, \beta) = (2, 18)$,

$$E\{p\} = \frac{1}{10}, \quad V\{p\} = \frac{3}{700}.$$

This implies that the failure probability at the first trial is approximately 0.1, on average, and the failure probability 0.1 of the second case is more certain than that of the first case because its variance 3/700 is smaller than 9/1100.

Table 6.2 presents the optimum retrial number N^* and the resulting mean time $\ell_1(N^*)/T_1$ for $T_2/T_1 = 0.1$–1.0. This indicates that N^* increases directly with T_2/T_1, and its values in the second case are approximately two times that of the first. When T_2/T_1 increases, we should attempt to do the trial by iterating retrials rather than by returning to the first after some time T_2. Furthermore, there is a tendency for $\ell_1(N^*)/T_1$ to increase slowly and converge to a limiting value as T_2 increases. This shows that the mean time does not increase as long as N^* is adopted, even if T_2 increases

The effect of the prior distribution on N^* is examined as follows: The value of p is more convincing in the second case than in the first. Consequently, after experiencing the same number of retrial failures, the certainty of such conviction in the second case is still higher, i.e., the value of p is underestimated, so that N^* becomes larger than that in the first case and is about two times that of the first. ∎

6.3 ARQ Models with Intermittent Faults

We consider three ARQ models with intermittent faults that are mainly employed in data transmissions to achieve high reliability of communication [133–137]. We want to transmit the data accurately and rapidly to a recipient. However, faults in a communication system sometimes occur intermittently. It is assumed in Model 1 that the system repeats normal and fault states alternately. The data transmission fails with probability p_j at the jth retransmission when the system is in a fault state. Faults in Model 2 are hidden and become permanent failure [133] when the duration of hidden faults exceeds a threshold level. The data transmissions fail with constant probability p for hidden fault and probability 1 for permanent failure. The data transmission in Model 3 fails with probability p_0 in a normal state with no fault, p_1 for hidden fault, and p_2 for permanent failure.

ARQ strategies are adopted when the data transmission due to faults fails. However, the data throughput decreases significantly if retransmissions are repeated without limitation. To keep the level of data throughput, when all numbers N of transmissions have failed, the system is inspected and maintained. We repeat the above procedure until the data transmission succeeds. We derive the mean times to success for such models and discuss optimum retransmission number N^* that minimizes them, using the results of Sect. 6.1. A hybrid ARQ scheme, that combines the policies of transmitting the data together with error-detecting and error-correcting codes and of retransmitting the same data when errors have been detected, was proposed and analyzed [136, 137].

6.3.1 Model 1

Faults occur intermittently, $i.e.$, the system with faults repeats the normal state (State 0) and fault state (State 1) alternately. The times in respective normal and fault states are independent and have identical exponential distributions $(1 - e^{-\lambda t})$ and $(1 - e^{-\mu t})$ for $\mu > \lambda$. The transitions probabilities $P_{ij}(t)$ $(i, j = 0, 1)$ that the system is in State j at time t when it starts in State i at time 0 are [1, p. 221]

$$P_{00}(t) = \frac{\mu}{\lambda + \mu} + \frac{\lambda}{\lambda + \mu} e^{-(\lambda+\mu)t},$$

$$P_{10}(t) = \frac{\mu}{\lambda + \mu} - \frac{\mu}{\lambda + \mu} e^{-(\lambda+\mu)t}, \tag{6.18}$$

and $P_{01}(t) = 1 - P_{00}(t)$, $P_{11}(t) = 1 - P_{10}(t)$.

It is assumed that when the system is in a normal state, $i.e.$, no fault occurs, the data transmission succeeds certainly, and when the system is in a fault state, it succeeds with probability q_j and fails with probability $p_j \equiv 1 - q_j$ at the jth $(j = 1, 2, \ldots, N)$ retransmissions with time T_1. When N

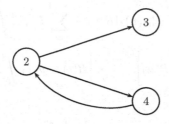

Fig. 6.3. State-transition diagram of Model 1

retransmissions have failed consecutively including the first one, the system is inspected and maintained preventively, and the same data transmission begins again after time T_2.

We define the following states of the above data transmission:

State 2: The jth $(j = 1, 2, \ldots, N)$ retransmission begins.
State 3: The transmission succeeds.
State 4: All N retransmissions fail and the system is maintained.

The above states form a Markov renewal process [1, p. 30], where State 3 is an absorbing state, and both States 2 and 4 are regeneration points. The mass functions $Q_{ij}(t)$ $(i, j = 2, 3, 4)$ are derived as follows: Using the same notations in **(3)** of Sect. 6.1, the probability that the data transmission succeeds at the first transmission up to time t, when the system is in State 0 at time 0, is

$$\int_0^t P_{00}(u) \, dR(u) + q_1 \int_0^t P_{01}(u) \, dR(u) = R(t) - p_1 \int_0^t P_{01}(u) \, dR(u),$$

and the probability that it succeeds at the $(j + 1)$th retransmission is

$$P(j) \left[\int_0^t P_{01}(u) \, dR(u) \right] * \left[\int_0^t P_{11}(u) \, dR(u) \right]^{(j-1)}$$
$$* \left[\int_0^t P_{10}(u) \, dR(u) + q_{j+1} \int_0^t P_{11}(u) \, dR(u) \right]$$
$$= P(j) \left[\int_0^t P_{01}(u) \, dR(u) \right] * \left[\int_0^t P_{11}(u) \, dR(u) \right]^{(j-1)}$$
$$* \left[R(t) - p_{j+1} \int_0^t P_{11}(u) \, dR(u) \right],$$

where $P(j) \equiv p_1 p_2 \ldots p_j$ $(j = 1, 2, \ldots, N)$, $P(0) \equiv 1$, the asterisk denotes the pairwise Stieltjes convolution, and $[\Phi(t)]^{(j)}$ denotes the j-fold Stieltjes convolution of $\Phi(t)$ with itself, i.e., $[\Phi(t)]^{(j)} \equiv \int_0^t \Phi^{(j-1)}(t - u) d\Phi(u)$, $[\Phi(t)]^{(0)} \equiv 1$ and $a(t) * b(t) \equiv \int_0^t a(t - u) db(u)$. Thus,

$$Q_{23}(t) = R(t) - p_1 \int_0^t P_{01}(u)\, dR(u) + \sum_{j=1}^{N-1} P(j) \left[\int_0^t P_{01}(u)\, dR(u) \right]$$

$$* \left[\int_0^t P_{11}(u)\, dR(u) \right]^{(j-1)} * \left[R(t) - p_{j+1} \int_0^t P_{11}(u)\, dR(u) \right]. \quad (6.19)$$

Similarly,

$$Q_{24}(t) = P(N) \left[\int_0^t P_{01}(u)\, dR(u) \right] * \left[\int_0^t P_{11}(u)\, dR(u) \right]^{(N-1)}, \quad (6.20)$$

$$Q_{42}(t) = G(t). \quad (6.21)$$

Let $H_{23}(t)$ be the first-passage time distribution from State 2 to State 3. Then, we have a renewal equation:

$$H_{23}(t) = Q_{23}(t) + Q_{24}(t) * Q_{42}(t) * H_{23}(t). \quad (6.22)$$

Thus, transforming the LS transform of (6.22) and solving it for $H_{23}(t)$,

$$H_{23}^*(s) = \frac{Q_{23}^*(s)}{1 - Q_{24}^*(s)Q_{42}^*(s)}. \quad (6.23)$$

Substituting $Q_{ij}^*(s)$ in (6.19) – (6.21) in (6.23),

$$H_{23}^*(s) = \frac{\begin{array}{l} R^*(s) - p_1 \int_0^\infty e^{-st} P_{01}(t) dR(t) \\ + \int_0^\infty e^{-st} P_{01}(t) dR(t) \sum_{j=1}^{N-1} P(j) \\ \times [\int_0^\infty e^{-st} P_{11}(t) dR(t)]^{j-1} [R^*(s) - p_{j+1} \int_0^\infty e^{-st} P_{11}(t) dR(t)] \end{array}}{1 - G^*(s) P(N) \int_0^\infty e^{-st} P_{01}(t) dR(t) [\int_0^\infty e^{-st} P_{11}(t) dR(t)]^{N-1}}. \quad (6.24)$$

Therefore, the mean time from State 2 to State 3 is

$$\ell_1(N) \equiv \lim_{s \to 0} \frac{1 - H_{23}^*(s)}{s}$$

$$= \frac{T_1 + T_2 + T_1 \int_0^\infty P_{01}(t) dR(t) \sum_{j=1}^{N-1} P(j) [\int_0^\infty P_{11}(t) dR(t)]^{j-1}}{1 - P(N) \int_0^\infty P_{01}(t) dR(t) [\int_0^\infty P_{11}(t) dR(t)]^{N-1}} - T_2$$

$$(N = 1, 2, \dots). \quad (6.25)$$

Suppose that both retransmission and maintenance times from State 4 to State 2 are constant, i.e., $R(t) \equiv 1$ for $t \geq T_1$, 0 for $t < T_1$ and $G(t) \equiv 1$ for $t \geq T_2$, 0 for $t < T_2$. Then, $\int_0^\infty P_{ij}(t) dR(t) = P_{ij}(T_1)$. We find an optimum number N^* that minimizes $\ell_1(N)$. From the inequality $\ell_1(N+1) - \ell_1(N) \geq 0$,

$$\frac{1 - P(N) P_{01}(T_1) [P_{11}(T_1)]^{N-1}}{1 - p_{N+1} P_{11}(T_1)} - P_{01}(T_1) \sum_{j=1}^{N-1} P(j) [P_{11}(T_1)]^{j-1}$$

$$\geq \frac{T_1 + T_2}{T_1} \quad (N = 1, 2, \dots), \quad (6.26)$$

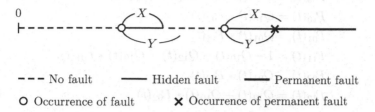

Fig. 6.4. Process of intermittent faults

where $\sum_1^0 \equiv 0$. It is easily proved that if p_j increases strictly with j, then the left-hand side of (6.26) also increases strictly with N. Therefore, if there exists a finite N such that (6.26), then an optimum N^* is given by a unique minimum that satisfies (6.26). In particular, when $P_{01}(T_1) = P_{11}(T_1) = 1$, (6.26) agrees with (6.4).

6.3.2 Model 2

Faults occur in a communication system according to an exponential distribution $(1 - e^{-\lambda t})$ and are hidden. When the duration X in hidden faults exceeds an upper limit time Y, faults become permanent failures, and otherwise, they get out of a hidden state, that is, if the event $\{X \le Y\}$ occurs, then hidden faults disappear, and conversely, if the event $\{X > Y\}$ occurs, then they become permanent failure (Fig. 6.4). It is assumed that both random variables X and Y are independent and have exponential distributions $\Pr\{X \le t\} = 1 - e^{-\mu t}$ and $\Pr\{Y \le t\} = 1 - e^{-\theta t}$, respectively.

We define the following states of intermittent faults [138]:

State 0: No fault occurs and the system is in a normal condition.

State 1: Hidden fault occurs.

State 2: Permanent failure occurs.

By the method similar to Model 1, we have the following mass functions $Q_{ij}(t)$ $(i = 0, 1, j = 0, 1, 2)$ from State i to State j in time t:

$$Q_{01}(t) = 1 - e^{-\lambda t},$$

$$Q_{10}(t) = \int_0^t e^{-\theta u} \mu e^{-\mu u}\, du = \frac{\mu}{\mu + \theta}[1 - e^{-(\mu+\theta)t}],$$

$$Q_{12}(t) = \int_0^t e^{-\mu u} \theta e^{-\theta u}\, du = \frac{\theta}{\mu + \theta}[1 - e^{-(\mu+\theta)t}]. \qquad (6.27)$$

Next, let $P_{ij}(t)$ denote the transition probabilities from State i at time 0 to State j at time t. Then, we have the equations:

$$P_{00}(t) = 1 - Q_{01}(t) + Q_{01}(t) * P_{10}(t),$$
$$P_{10}(t) = Q_{10}(t) * P_{00}(t),$$
$$P_{01}(t) = Q_{01}(t) * P_{11}(t),$$
$$P_{11}(t) = 1 - Q_{10}(t) - Q_{12}(t) + Q_{10}(t) * P_{01}(t),$$
$$P_{02}(t) = Q_{01}(t) * P_{12}(t),$$
$$P_{12}(t) = Q_{12}(t) + Q_{10}(t) * P_{02}(t). \tag{6.28}$$

Forming the LS transforms of (6.28) and rearranging them,

$$P_{00}^*(s) = \frac{1 - Q_{01}^*(s)}{1 - Q_{01}^*(s)Q_{10}^*(s)},$$
$$P_{01}^*(s) = \frac{Q_{01}^*(s)[1 - Q_{10}^*(s) - Q_{12}^*(s)]}{1 - Q_{01}^*(s)Q_{10}^*(s)},$$
$$P_{10}^*(s) = \frac{Q_{10}^*(s)[1 - Q_{01}^*(s)]}{1 - Q_{01}^*(s)Q_{10}^*(s)},$$
$$P_{11}^*(s) = \frac{1 - Q_{10}^*(s) - Q_{12}^*(s)}{1 - Q_{01}^*(s)Q_{10}^*(s)}. \tag{6.29}$$

Thus, substituting (6.27) in (6.29) and taking the inverse LS transforms, the transition probabilities from State i to State j $(i, j = 0, 1, 2)$ are

$$P_{00}(t) = \frac{1}{\gamma_1 - \gamma_2}[(\mu + \theta - \gamma_2)e^{-\gamma_2 t} - (\mu + \theta - \gamma_1)e^{-\gamma_1 t}],$$
$$P_{01}(t) = \frac{\lambda}{\gamma_1 - \gamma_2}(e^{-\gamma_2 t} - e^{-\gamma_1 t}),$$
$$P_{10}(t) = \frac{\mu}{\gamma_1 - \gamma_2}(e^{-\gamma_2 t} - e^{-\gamma_1 t}),$$
$$P_{11}(t) = \frac{1}{\gamma_1 - \gamma_2}[(\lambda - \gamma_2)e^{-\gamma_2 t} - (\lambda - \gamma_1)e^{-\gamma_1 t}], \tag{6.30}$$

$P_{02}(t) = 1 - P_{00}(t) - P_{01}(t)$, and $P_{12}(t) = 1 - P_{10}(t) - P_{11}(t)$, where

$$\gamma_1 \equiv \frac{1}{2}[\lambda + \mu + \theta + \sqrt{(\lambda + \mu + \theta)^2 - 4\lambda\theta}],$$
$$\gamma_2 \equiv \frac{1}{2}[\lambda + \mu + \theta - \sqrt{(\lambda + \mu + \theta)^2 - 4\lambda\theta}].$$

It is assumed that when the system is in State 0, the data transmission succeeds with probability 1, when the system is in State 1, it succeeds with probability q and fails with probability $p \equiv 1 - q$, and when the system is in State 2, it fails with probability 1. In addition, distributions $R(t)$ and $G(t)$ are degenerate distributions placing unit masses at times T_1 and T_2, respectively. The other assumptions are the same as those of Model 1. Suppose that the system is in State 0 at time 0. Then, the probability that the data transmission succeeds until the Nth number is

$$\overline{P}(N) = 1 - pP_{01}(T_1) - P_{02}(T_1) + pP_{01}(T_1)[1 - pP_{11}(T_1) - P_{12}(T_1)] \sum_{j=0}^{N-2} [pP_{11}(T_1)]^j,$$

and the probability that all N retransmissions have failed is

$$P(N) = P_{02}(T_1) + pP_{01}(T_1)\left\{ P_{12}(T_1) \sum_{j=0}^{N-2} [pP_{11}(T_1)]^j + [pP_{11}(T_1)]^{N-1} \right\},$$

where $\overline{P}(N) + P(N) = 1$.

We call the time from the beginning of transmission to N failures or success *one period*. Then, the mean time of one period is

$$\begin{aligned}\ell_1 = T_1 \Bigg\{ & 1 - pP_{01}(T_1) - P_{02}(T_1) + pP_{01}(T_1)[1 - pP_{11}(T_1) - P_{12}(T_1)] \\ & \times \sum_{j=0}^{N-2} (j+2)[pP_{11}(T_1)]^j + NP(N) \Bigg\} \\ = T_1 \Bigg\{ & 1 + (N-1)\left[P_{02}(T_1) + \frac{pP_{01}(T_1)P_{12}(T_1)}{1 - pP_{11}(T_1)} \right] \\ & + pP_{01}(T_1) \frac{1 - pP_{11}(T_1) - P_{12}(T_1)}{1 - pP_{11}(T_1)} \frac{1 - [pP_{11}(T_1)]^{N-1}}{1 - pP_{11}(T_1)} \Bigg\}.\end{aligned}$$

Therefore, the mean time in which the data transmission succeeds is

$$\ell_2(N) = \ell_1 + P(N)[T_2 + \ell_2(N)],$$

i.e.,

$$\begin{aligned}\ell_2(N) &= \frac{\ell_1 + T_2 P(N)}{\overline{P}(N)} \\ &= \frac{T_1}{1 - pP_{11}(T_1)} - T_2 \\ &+ \frac{\begin{array}{l} T_1 + T_2 - T_1\{[P_{00}(T_1) + qP_{01}(T_1)]/[1 - pP_{11}(T_1)]\} \\ +(N-1)T_1\{P_{02}(T_1) + pP_{01}(T_1)P_{12}(T_1)/[1 - pP_{11}(T_1)]\} \end{array}}{\begin{array}{l} P_{00}(T_1) + qP_{01}(T_1) \\ + pP_{01}(T_1)[P_{10}(T_1) + qP_{11}(T_1)]\{1 - [pP_{11}(T_1)]^{N-1}\}/[1 - pP_{11}(T_1)] \end{array}} \\ & \qquad (N = 1, 2, \ldots). \quad (6.31)\end{aligned}$$

We find an optimum number N^* that minimizes $\ell_2(N)$. Using the following notations:

$$A \equiv T_1 + T_2 - \frac{T_1[P_{00}(T_1) + qP_{01}(T_1)]}{1 - pP_{11}(T_1)},$$

$$B \equiv T_1 \left[P_{02}(T_1) + \frac{pP_{01}(T_1)P_{12}(T_1)}{1 - pP_{11}(T_1)} \right],$$

$$C \equiv P_{00}(T_1) + qP_{01}(T_1),$$

$$D \equiv pP_{01}(T_1)\frac{P_{10}(T_1) + qP_{11}(T_1)}{1 - pP_{11}(T_1)},$$

the mean time $\ell_2(N)$ in (6.31) is rewritten as

$$\ell_2(N) = \frac{T_1}{1 - pP_{11}(T_1)} - T_2$$
$$+ \frac{A + (N-1)B}{C + D\{1 - [pP_{11}(T_1)]^{N-1}\}} \qquad (N = 1, 2, \dots). \qquad (6.32)$$

It is clearly seen that a finite N^* $(1 \le N^* < \infty)$ exists because $\lim_{N \to \infty} \ell_2(N) = \infty$. From the inequality $\ell_2(N+1) - \ell_2(N) \ge 0$,

$$\frac{1 + C/D}{[1 - pP_{11}(T_1)][pP_{11}(T_1)]^{N-1}} - (N-1) \ge \frac{A}{B} + \frac{1}{1 - pP_{11}(T_1)}$$
$$(N = 1, 2, \dots). \qquad (6.33)$$

Letting $L(N)$ be the the left-hand side of (6.33), we easily have that $L(\infty) \equiv \lim_{N \to \infty} L(N) = \infty$,

$$L(1) = \frac{1 + C/D}{1 - pP_{11}(T_1)},$$

$$L(N+1) - L(N) = \frac{1 + C/D - [pP_{11}(T_1)]^N}{[pP_{11}(T_1)]^N} > 0.$$

Therefore, an optimum N^* $(1 \le N^* < \infty)$ is given by a finite and unique minimum that satisfies (6.33). If $C/D \ge (A/B)[1 - pP_{11}(T_1)]$, then $N^* = 1$.

6.3.3 Model 3

It is assumed in Model 2 that when the system is in State i $(i = 0, 1, 2)$, the data transmission succeeds with probability q_i and fails with probability $p_i \equiv 1 - q_i$, where $0 \le p_0 \le p_1 \le p_2 \le 1$. The other notations are the same as those of Model 2.

Let Q_{ij} be the probability that when the system is in State i $(i = 0, 1)$, the data transmission has failed at j times $(j = 1, 2, \dots, N)$ consecutively. Then, using (6.30), we have the following equations related to Q_{ij}:

$$Q_{0j} = P_{00}(T_1)p_0Q_{0j-1} + P_{01}(T_1)p_1Q_{1j-1} + P_{02}(T_1)p_2^j,$$
$$Q_{1j} = P_{10}(T_1)p_0Q_{0j-1} + P_{11}(T_1)p_1Q_{1j-1} + P_{12}(T_1)p_2^j, \qquad (6.34)$$

where $Q_{i0} \equiv 1$. To obtain Q_{ij} explicitly, we introduce the notation of the generating function as

$$Q_i^*(z) \equiv \sum_{j=0}^{\infty} Q_{ij} z^j \qquad (i = 0, 1)$$

for $|z| \leq 1$. Thus, from (6.34),

$$Q_0^*(z) = 1 + P_{00}(T_1)p_0 z Q_0^*(z) + P_{01}(T_1)p_1 z Q_1^*(z) + P_{02}(T_1)\frac{p_2 z}{1 - p_2 z},$$

$$Q_1^*(z) = 1 + P_{10}(T_1)p_0 z Q_0^*(z) + P_{11}(T_1)p_1 z Q_1^*(z) + P_{12}(T_1)\frac{p_2 z}{1 - p_2 z}.$$

Solving the above equations for $Q_0^*(z)$,

$$Q_0^*(z) = \frac{\begin{aligned}&1 - p_1 z[P_{11}(T_1) - P_{01}(T_1)] \\ &+ p_2 z\{P_{02}(T_1) - p_1 z[P_{11}(T_1)P_{02}(T_1) - P_{01}(T_1)P_{12}(T_1)]\}/(1 - p_2 z)\end{aligned}}{[1 - p_0 z P_{00}(T_1)][1 - p_1 z P_{11}(T_1)] - p_0 p_1 z^2 P_{01}(T_1) P_{10}(T_1)}.$$
$$(6.35)$$

Furthermore, expanding $Q_0^*(z)$ into z^j,

$$Q_0^*(z) = \frac{a_1 - p_1[P_{11}(T_1) - P_{01}(T_1)]}{(a_1 - a_2)(1 - a_1 z)} + \frac{a_2 - p_1[P_{11}(T_1) - P_{01}(T_1)]}{(a_2 - a_1)(1 - a_2 z)}$$
$$+ \frac{p_2 z\{a_1 P_{02}(T_1) - p_1[P_{11}(T_1)P_{02}(T_1) - P_{01}(T_1)P_{12}(T_1)]\}/(1 - p_2 z)}{(a_1 - a_2)(a_1 - p_2)(1 - a_1 z)}$$
$$+ \frac{p_2 z\{a_2 P_{02}(T_1) - p_1[P_{11}(T_1)P_{02}(T_1) - P_{01}(T_1)P_{12}(T_1)]\}/(1 - p_2 z)}{(a_2 - a_1)(a_2 - p_2)(1 - a_2 z)}$$
$$(6.36)$$

except that $p_0 = p_1 = 0$, where

$$a_1 \equiv \frac{1}{2}\Big\{ p_0 P_{00}(T_1) + p_1 P_{11}(T_1)$$
$$+ \sqrt{[p_0 P_{00}(T_1) - p_1 P_{11}(T_1)]^2 + 4 p_0 p_1 P_{01}(T_1) P_{10}(T_1)} \Big\},$$

$$a_2 \equiv \frac{1}{2}\Big\{ p_0 P_{00}(T_1) + p_1 P_{11}(T_1)$$
$$- \sqrt{[p_0 P_{00}(T_1) - p_1 P_{11}(T_1)]^2 + 4 p_0 p_1 P_{01}(T_1) P_{10}(T_1)} \Big\}.$$

It is easily noted that $a_1 > a_2 > 0$. Thus, from the definition of $Q_0^*(z)$, and (6.36),

$$Q_{0j} = \frac{1}{a_1 - a_2}$$

$$\times \left(\{a_1 - p_1[P_{11}(T_1) - P_{01}(T_1)]\}a_1^j - \{a_2 - p_1[P_{11}(T_1) - P_{01}(T_1)]\}a_2^j \right)$$

$$+ \frac{p_2}{a_1 - a_2}$$

$$\times \left(\{a_1 P_{02}(T_1) - p_1[P_{11}(T_1)P_{02}(T_1) - P_{01}(T_1)P_{12}(T_1)]\}\frac{a_1^j - p_2^j}{a_1 - p_2} \right.$$

$$\left. - \{a_2 P_{02}(T_1) - p_1[P_{11}(T_1)P_{02}(T_1) - P_{01}(T_1)P_{12}(T_1)]\}\frac{a_2^j - p_2^j}{a_2 - p_2} \right)$$

$$(j = 0, 1, 2, \ldots, N). \tag{6.37}$$

It is clearly seen that $Q_{00} = 1$ and

$$Q_{01} = p_0 P_{00}(T_1) + p_1 P_{01}(T_1) + p_2 P_{02}(T_1).$$

Because the probability that the data transmission at the jth retransmission is $Q_{0j-1} - Q_{0j}$, and the probability that N retransmissions have failed is Q_{0N}, the mean time of one period of retransmissions is

$$\ell_1 = T_1 \left[\sum_{j=1}^{N} j(Q_{0j-1} - Q_{0j}) + NQ_{0N} \right] = T_1 \sum_{j=0}^{N-1} Q_{0j}.$$

Thus, the mean time in which the data transmission succeeds is, from (6.31),

$$\ell_3(N) = \frac{T_1 \sum_{j=0}^{N-1} Q_{0j} + T_2 Q_{0N}}{1 - Q_{0N}} \qquad (N = 1, 2, \ldots). \tag{6.38}$$

We discuss an optimum number N^* that minimizes $\ell_3(N)$. From the inequality $\ell_3(N+1) - \ell_3(N) \geq 0$,

$$\frac{Q_{0N}}{Q_{0N} - Q_{0N+1}}(1 - Q_{0N}) - \sum_{j=0}^{N-1} Q_{0j} \geq \frac{T_2}{T_1} \qquad (N = 1, 2, \ldots). \tag{6.39}$$

Letting $L(N)$ be the right-hand side of (6.39),

$$L(N+1) - L(N) = (1 - Q_{0N+1}) \left(\frac{Q_{0N+1}}{Q_{0N+1} - Q_{0N+2}} - \frac{Q_{0N}}{Q_{0N} - Q_{0N+1}} \right).$$

Thus, if $Q(N) \equiv Q_{0N+1}/Q_{0N}$ increases strictly, then $L(N)$ also increases strictly. In this case,

$$L(N) > \frac{Q_{0N}}{Q_{0N} - Q_{0N+1}}(1 - Q_{01}) - 1 \qquad (N = 2, 3, \ldots),$$

because

$$\frac{Q_{0N}}{Q_{0N} - Q_{0N+1}}(1 - Q_{0N}) - \sum_{j=0}^{N-1} Q_{0j} - \frac{Q_{0N}}{Q_{0N} - Q_{0N+1}}(1 - Q_{01}) + 1$$

$$= \frac{Q_{0N}}{Q_{0N} - Q_{0N+1}}(Q_{01} - Q_{0N}) - \sum_{j=1}^{N-1} Q_{0j}$$

$$= \frac{1}{Q_{0N} - Q_{0N+1}}\left(Q_{0N+1}\sum_{j=1}^{N-1} Q_{0j} - Q_{0N}\sum_{j=1}^{N-1} Q_{0j+1}\right)$$

$$> \frac{1}{Q_{0N} - Q_{0N+1}}\left(Q_{0N+1}\sum_{j=1}^{N-1} Q_{0j} - Q_{0N}\frac{Q_{0N+1}}{Q_{0N}}\sum_{j=1}^{N-1} Q_{0j}\right) = 0.$$

From the above discussions, when $Q(N)$ increases strictly and $Q(\infty) \equiv \lim_{N\to\infty} Q(N)$, if

$$\frac{1}{1 - Q(\infty)}(1 - Q_{01}) \geq \frac{T_1 + T_2}{T_1},$$

i.e.,

$$Q(\infty) \geq 1 - \frac{T_1}{T_1 + T_2}(1 - Q_{01}),$$

then there exists a finite and unique minimum N^* ($1 \leq N^* < \infty$) that satisfies (6.39).

In particular, when $p_2 > a_1$, $Q(\infty) = p_2$, and hence, if

$$q_2 \leq \frac{T_1}{T_1 + T_2}[1 - p_0 P_{00}(T_1) - p_1 P_{01}(T_1) - p_2 P_{02}(T_1)],$$

then a finite N^* exists uniquely. Furthermore, when T_1 is very small, we easily have that $P_{00}(T_1)P_{11}(T_1) > P_{01}(T_1)P_{10}(T_1)$, so that from the definition of a_1, it is clearly seen that $a_1 < p_0 P_{00}(T_1) + p_1 P_{11}(T_1)$. Thus, if $p_2 \geq p_0 P_{00}(T_1) + p_1 P_{11}(T_1)$, then $p_2 > a_1$.

Finally, consider the particular case of $p_0 = p_1 = 0$ and $0 < p_2 < 1$. Then, the mean time $\ell_3(N)$ in (6.38) becomes simply

$$\ell_3(N) = \frac{T_2 - K}{1 - P_{02}(T_1)p_2^N} + \frac{T_1}{q_2} - T_2, \tag{6.40}$$

where $K \equiv T_1 p_2[1 - P_{02}(T_1)]/q_2$. We have the following results:

(i) If $T_2 \geq K$, then $\ell_3(N)$ decreases with N, so that $N^* = \infty$ and

$$\ell_3(\infty) \equiv \lim_{N\to\infty} \ell_3(N) = T_1\left[1 + \frac{P_{02}(T_1)p_2}{q_2}\right]. \tag{6.41}$$

(ii) If $T_2 < K$, then $\ell_3(N)$ increases with N, so that $N^* = 1$ and

$$\ell_3(1) = \frac{T_1 + T_2 P_{02}(T_1)p_2}{1 - P_{02}(T_1)p_2}. \tag{6.42}$$

Furthermore, when $p_2 = 1$, the mean time to success is

$$\ell_3(N) = \frac{T_1[1 + (N-1)P_{02}(T_1)] + T_2}{1 - P_{02}(T_1)} - T_2. \tag{6.43}$$

Thus, $\ell_3(N)$ increases with N, so that $N^* = 1$. Therefore, when $p_0 = p_1 = 0$, there exists no finite N^* ($N^* \geq 2$) that minimizes $l_3(N)$ in (6.38).

Example 6.3.   We set time T_1 required for one transmission as a unit time. Table 6.3 presents the optimum numbers N^* for p_1 when $1/\lambda = 14,400T_1, 21,600T_1, 1/\mu = 300T_1, 1/\theta = 60T_1, 300T_1, T_2 = 60T_1, 90T_1$, and $p_0 = 0.1, p_2 = 0.99$. For example, when $T_1 = 1$ second, $1/\lambda = 21,600T_1 = 6$ hours, $1/\mu = 1/\theta = 300T_1 = 5$ minutes, and $T_2 = 60T_1 = 1$ minute. This shows the change of the optimum N^*.

Suppose that the error rate per bit is about 10^{-5}, and the data length is about 10^4 under normal conditions, *i.e.*, $p_0 = 0.1$. Thus, the error rate due to intermittent faults increases from p_0 to $p_1 = 0.3$–0.7.

Table 6.3 shows the change of N^* for $1/\lambda$, $1/\theta$, T_2, and p_1. This indicates that N^* increases with T_2, $1/\theta$, and p_1. However, the optimum numbers are changed little by $1/\lambda$ and are determined approximately by p_i ($i = 0, 1, 2$), T_2, and $1/\theta$.

Tables 6.4 and 6.5 present N^* for p_i when $1/\lambda = 21,600T_1$, $1/\mu = 300T_1$, $T_2 = 60T_1$, and $1/\theta = 300T_1$. The values of N^* increases with p_0 for a fixed p_2 in Table 6.4 and decreases with p_2 for a fixed p_0, *i.e.*, N^* increases with p_0 and p_1, and conversely, decreases with p_2. These results show that N^* increases with p_0 because the number of failure transmissions in State 0, that seems to be originally independent of N, increases. Conversely, N^* decreases with p_2 because the probability of failures in State 2 increases. ∎

Table 6.3. Optimum number N^* when $p_0 = 0.1$ and $p_2 = 0.99$, and $1/\mu = 300T_1$

$1/\lambda$	p_1	$T_2 = 60T_1$ $1/\theta$		$T_2 = 90T_1$ $1/\theta$	
		$60\ T_1$	$300\ T_1$	$60\ T_1$	$300\ T_1$
	0.3	9	10	10	11
	0.4	10	12	12	14
$14,400\ T_1$	0.5	12	15	15	17
	0.6	15	19	19	22
	0.7	19	25	25	30
	0.3	9	10	10	11
	0.4	10	12	12	14
$21,600\ T_1$	0.5	12	15	15	17
	0.6	15	19	19	22
	0.7	19	25	25	30

Table 6.4. Optimum number N^* when $1/\mu = 1/\theta = 300T_1$, $1/\lambda = 21,600T_1$, and $T_2 = 60T_1$

p_1	$p_2 = 0.99$ p_0					
	0.0	0.1	0.2	0.3	0.4	0.5
0.1	5	10	-	-	-	-
0.2	7	10	13	-	-	-
0.3	9	10	13	17	-	-
0.4	12	12	13	17	22	-
0.5	14	15	15	17	22	28

p_1	$p_2 = 1.0$ p_0					
	0.0	0.1	0.2	0.3	0.4	0.5
0.1	5	9	-	-	-	-
0.2	7	9	13	-	-	-
0.3	8	9	13	16	-	-
0.4	10	11	13	16	21	-
0.5	13	13	14	16	21	26

Table 6.5. Optimum number N^* when $1/\mu = 1/\theta = 300T_1$, $1/\lambda = 21, 600T_1$, and $T_2 = 60T_1$

p_1	$p_0 = 0.0$			$p_0 = 0.1$		
	p_2			p_2		
	0.99	0.999	1.0	0.99	0.999	1.0
0.1	5	5	5	10	9	9
0.2	7	7	7	10	9	9
0.3	9	8	8	10	9	9
0.4	12	10	10	12	11	11
0.5	14	13	13	15	13	13

7

Optimum Checkpoint Intervals for Fault Detection

Computer systems have been urgently required to operate normally and effectively as communication and information systems have been developed rapidly and are remarkably complicated. However, some errors in systems often occur due to noises, human errors, software bugs, hardware faults, computer viruses, and so on, and last, they might become faults and incur their failures. To protect against such faults, various fault tolerant techniques such as the redundancy of processors and memories and the configuration of systems have been provided [10–13]. The high reliability and effective performance of real systems could be achieved by the use of redundant techniques, as shown in Sect. 1.2.

Partial data loss and operational errors in fault tolerant computing are generally called *error* and *fault* caused by errors. *Failure* indicates that faults are recognized on the exterior of systems [10]. Some faults due to operational errors may be detected after some time has passed, and system consistency may be lost by them. Then, we should restore a consistent state just before fault occurrences by some recovery techniques. The operation of taking copies of the normal state is called *checkpoint* [58, 139, 140]. When faults have occurred, the process goes back to the nearest checkpoint time by *rollback operation*, and its retry is done, using a consistent state stored in the checkpoint time.

Several studies for deciding optimum checkpoint frequencies have been done: The performance and reliability of a double modular system with one spare module were evaluated [141, 142]. Furthermore, the performance of checkpoint schemes with task duplication was evaluated [60, 61]. The optimum instruction-retry period that minimizes the probability of the dynamic failure by a triple modular controller was derived [143]. Evaluation models with finite checkpoints and bounded rollback were discussed [144]. Recently, most systems consist of distributed systems as computer network technologies have developed rapidly. A general model of distributed systems is a mobile network system. Coordinated and uncoordinated protocols to achieve checkpointing in such distributed systems were introduced and evaluated [10, 144].

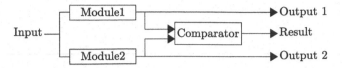

Fig. 7.1. Error detection by a double modular system

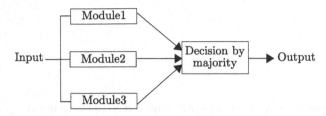

Fig. 7.2. Error masking by a triple modular system

As application examples of results in Chaps 2–5, we take up the following checkpoint models of computer systems and analyze them from the viewpoint of reliability: Suppose that we have to complete the process of one task with a finite execution time S. Modules are elements such as logical circuits or processors that execute certain lumped parts of the task.

In Sect. 7.1, we first adopt a double modular system as a redundant technique that executes the process of one task. In this system, two modules are functionally equivalent, their states are compared, and errors may be detected. Introducing the overhead to store and compare the states, the total mean time to complete the process of one task is obtained, and an optimum checkpoint time that minimizes it is derived, using the partition method in Chap. 3 [145]. Furthermore, we adopt a triple modular system and a majority decision system as error masking techniques [146].

In Sect. 7.2, it is assumed that error rates during checkpoint intervals increase with their numbers for a double modular system. The mean time to complete the process of one task is obtained, and optimum sequential checkpoint times are computed numerically, using the techniques used in Chap. 4 [147]. Furthermore, one approximation method for obtaining optimum times and a checkpoint model with a general error rate are proposed.

In Sect. 7.3, we consider two extended checkpoint models with one spare module [148] and three detection schemes for a double modular system [149]. The mean times to complete the process for each model are obtained. The optimum checkpoint times are computed numerically and compared with those of the standard model in Sect. 7.1

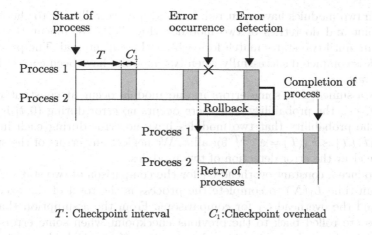

Fig. 7.3. Sample of the execution of a double modular system

7.1 Checkpoint Intervals of Redundant Systems

Suppose that a modular system executes the process of one task. We take up a double modular redundancy (Fig. 7.1) and a triple modular redundancy (Fig. 7.2) as redundant techniques for error detection and error masking [145]: When a native execution time S of one task is given, we divide it equally into time intervals. Introducing two overheads for comparison equally by duplication and decision by majority, we obtain the mean times to complete the processes successfully for two systems. Using the partition method in Sect. 3.1, we derive optimum checkpoint intervals that minimize the mean times.

Furthermore, we consider a redundant system of a majority decision with $(2n+1)$ modules as an error masking system, $i.e.$, an $(n+1)$-out-of-$(2n+1)$ system $(n = 1, 2, \dots)$ [146]. We compute the mean time to complete the process and decide numerically what majority system is optimum.

(1) Double Modular System

Suppose that S $(0 < S < \infty)$ is a native execution time of the process that does not include the overheads of retries and checkpoint generations. Then, we divide S equally into N $(N = 1, 2, \dots)$ time intervals and place checkpoints at every planned time T, where $T \equiv S/N$ (Fig. 3.1). Such checkpoints have two functions that store and compare the state of the process. To detect some errors in the process, we execute two independent modules and compare two states of modules at periodic checkpoint times kT $(k = 1, 2, \dots, N)$. If two states match equally, then two modules are correct, and we proceed to the next interval. Conversely, if two states do not match, then it is judged that some errors in two modules have occurred. In this case, we roll back to the newest checkpoint and do retries of two modules (Fig. 7.3). We repeat the above procedure until two states match for each checkpoint interval. The process of one task is completed successfully when two modules have been correct for all N intervals.

It is assumed that some errors in one module occur at constant rate λ ($\lambda > 0$), i.e., the probability that there occurs no error during $(0, t]$ is $e^{-\lambda t}$. Thus, the probability that two modules have no error during each interval $((k-1)T, kT]$ is $\overline{F}_1(T) = e^{-2\lambda T}$ for all k. We neglect any errors of the system to make clear the error detection of two modules.

Introduce a constant overhead C_1 for the comparison of two states. Then, the mean time $L_1(N)$ to complete the process is the total of the execution times and the overhead C_1 for comparisons. From the assumption that two modules are rolled back to the previous checkpoint when some errors have been detected at a checkpoint time, the mean time for each checkpoint interval $((k-1)T, kT]$ is given by a renewal equation:

$$L_1(1) = (T + C_1)\, e^{-2\lambda T} + [T + C_1 + L_1(1)]\,(1 - e^{-2\lambda T}). \qquad (7.1)$$

Solving (7.1) with respect to $L_1(1)$,

$$L_1(1) = \frac{T + C_1}{e^{-2\lambda T}}. \qquad (7.2)$$

Thus, the mean time to complete the process is

$$L_1(N) \equiv N L_1(1) = N(T + C_1)\, e^{2\lambda T}. \qquad (7.3)$$

Because $T = S/N$,

$$L_1(N) = (S + N C_1)\, e^{2\lambda S/N} \qquad (N = 1, 2, \dots). \qquad (7.4)$$

We find an optimum number N_1^* that minimizes $L_1(N)$ for a specified S and C_1. It is clearly seen that $\lim_{N \to \infty} L_1(N) = \infty$ and

$$L_1(1) = (S + C_1)\, e^{2\lambda S}. \qquad (7.5)$$

Thus, a finite N_1^* ($1 \le N_1^* < \infty$) exists.

Setting $T = S/N$ in (7.4) and rewriting it in terms of the function T,

$$L_1(T) = S\left(1 + \frac{C_1}{T}\right) e^{2\lambda T} \qquad (7.6)$$

for $0 < T \le S$. Clearly, $\lim_{T \to 0} L_1(T) = \infty$ and $L_1(S)$ is given in (7.5). Thus, there exists an optimum $\tilde{T}_1$ ($0 < \tilde{T}_1 \le S$) that minimizes $L_1(T)$ in (7.6). Differentiating $L_1(T)$ with respect to T and setting it equal to zero,

$$T^2 + C_1 T - \frac{C_1}{2\lambda} = 0. \qquad (7.7)$$

Solving (7.7) for T,

$$\widetilde{T}_1 = \frac{C_1}{2}\left(\sqrt{1 + \frac{2}{\lambda C_1}} - 1\right). \tag{7.8}$$

Therefore, we get the following optimum number N_1^*, using the partition method in Sect. 3.1:

(i) When $\widetilde{T}_1 < S$, we set $[S/\widetilde{T}_1] \equiv N$ and calculate $L_1(N)$ and $L_1(N+1)$ from (7.4). If $L_1(N) \le L_1(N+1)$, then $N_1^* = N$, and conversely, if $L_1(N) > L_1(N+1)$, then $N_1^* = N+1$.

(ii) When $\widetilde{T}_1 \ge S$, $N_1^* = 1$, i.e., we should generate no checkpoint, and the mean time is given in (7.5).

Note that $\widetilde{T}_1$ in (7.8) does not depend on S. Thus, if S is very large, is changed greatly, or is unclear, then we may adopt $\widetilde{T}_1$ as an approximate checkpoint time of T_1^*.

(2) Triple Majority Decision System

Consider a majority decision system with three modules, i.e., a 2-out-of-3 system, as an error masking system. If more than two states of three modules match equally, then the process in this interval is correct (Fig. 7.4), i.e., the system can mask a single error. Then, the probability that the process is correct during each interval $((k-1)T, kT]$ is

$$\overline{F}_2(T) = e^{-3\lambda T} + 3e^{-2\lambda T}(1 - e^{-\lambda T}). \tag{7.9}$$

Let C_2 be the overhead for the comparison of a majority decision in terms of three modules. By a similar method for obtaining (7.3), the mean time to complete the process is

$$L_2(N) = \frac{N(T + C_2)}{\overline{F}_2(T)} = \frac{S + NC_2}{3e^{-2\lambda T} - 2e^{-3\lambda T}} \quad (N = 1, 2, \ldots). \tag{7.10}$$

Setting $T \equiv S/N$ in (7.10),

$$L_2(T) = \frac{S(1 + C_2/T)}{3e^{-2\lambda T} - 2e^{-3\lambda T}} \tag{7.11}$$

for $0 < T \le S$. Clearly, $\lim_{T \to 0} L_2(T) = \infty$, and

$$L_2(S) = \frac{S + C_2}{3e^{-2\lambda S} - 2e^{-3\lambda S}}. \tag{7.12}$$

Thus, there exists an optimum $\widetilde{T}_2$ $(0 < \widetilde{T}_2 \le S)$ that minimizes $L_2(T)$ in (7.11). Differentiating $L_2(T)$ with respect to T and setting it equal to zero,

$$(e^{\lambda T} - 1)\left(T^2 + C_2 T - \frac{C_2}{2\lambda}\right) = \frac{C_2}{6\lambda}. \tag{7.13}$$

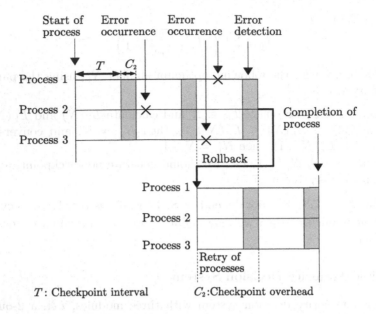

T: Checkpoint interval C_2:Checkpoint overhead

Fig. 7.4. Sample of the execution of triple processes

The left-hand side of (7.13) increases strictly from 0 to ∞, and hence, there exists a finite and unique $\widetilde{T}_2$ $(0 < \widetilde{T}_2 < \infty)$ that satisfies (7.13).

Therefore, using the partition method, we get an optimum number N_2^*. In the particular case of $C_1 = C_2$, $\widetilde{T}_2 > \widetilde{T}_1$, and hence, $N_2^* \leq N_1^*$.

(3) Majority Decision System

Consider a redundant system of a majority decision with $(2n + 1)$ modules as an error masking system, *i.e.*, an $(n+1)$-out-of-$(2n+1)$ system $(n = 1, 2, \dots)$. If more than $(n + 1)$ states of $(2n + 1)$ modules match equally, the process in this interval is correct. Then, the probability that the process is correct during $((k - 1)T, kT]$ is

$$\overline{F}_{n+1}(T) = \sum_{j=n+1}^{2n+1} \binom{2n + 1}{j} (e^{-\lambda T})^j (1 - e^{-\lambda T})^{2n+1-j}. \tag{7.14}$$

Thus, the mean time to complete the process is

$$L_{n+1}(N) = \frac{N(T + C_{n+1})}{\overline{F}_{n+1}(T)} \qquad (N = 1, 2, \dots), \tag{7.15}$$

where C_{n+1} is the overhead for the comparison of a majority decision in $(2n + 1)$ modules. When $n = 1$, $L_2(N)$ agrees with (7.10).

Table 7.1. Optimum checkpoint number N_1^*, interval λT_1^*, and mean time $\lambda L_1(N_1^*)$ for a double modular system when $\lambda S = 10^{-1}$

$\lambda C_1 \times 10^3$	$\lambda \widetilde{T}_1 \times 10^2$	N_1^*	$\lambda L_1(N_1^*) \times 10^2$	$\lambda T_1^* \times 10^2$
0.5	1.56	6	10.65	1.67
1.0	2.19	5	10.93	2.00
1.5	2.66	4	11.14	2.50
2.0	3.06	3	11.33	3.33
3.0	3.73	3	11.65	3.33
4.0	4.28	2	11.94	5.00
5.0	4.76	2	12.16	5.00
10.0	6.59	2	13.26	5.00
20.0	9.05	1	14.66	10.00
30.0	10.84	1	15.88	10.00

Table 7.2. Optimum checkpoint number N_2^*, interval λT_2^*, and mean time $\lambda L_2(N_2^*)$ for a triple modular system when $\lambda S = 10^{-1}$

$\lambda C_2 \times 10^3$	$\lambda \widetilde{T}_2 \times 10^2$	N_2^*	$\lambda L_2(N_2^*) \times 10^2$	$\lambda T_2^* \times 10^2$
0.1	2.61	4	10.06	2.50
0.2	3.30	3	10.09	3.33
0.3	3.79	3	10.12	3.33
0.4	4.18	2	10.15	5.00
0.5	4.51	2	10.17	5.00
1.0	5.72	2	10.27	5.00
1.5	6.58	2	10.37	5.00
2.0	7.27	1	10.47	10.00
3.0	8.36	1	10.57	10.00
4.0	9.23	1	10.67	10.00
5.0	9.97	1	10.77	10.00
10.0	12.68	1	11.29	10.00
20.0	16.09	1	12.31	10.00
30.0	18.47	1	13.34	10.00

Table 7.3. Optimum checkpoint number N_{n+1}^* and mean time $\lambda L_{n+1}(N_{n+1}^*)$ for an $(n+1)$-out-of-$(2n+1)$ system when $\lambda S = 10^{-1}$

	$\lambda C_1 = 0.1 \times 10^{-3}$		$\lambda C_1 = 0.5 \times 10^{-3}$	
n	N_{n+1}^*	$\lambda L_{n+1}(N_{n+1}^*) \times 10^2$	N_{n+1}^*	$\lambda L_{n+1}(N_{n+1}^*) \times 10^2$
1	3	10.12	2	10.37
2	1	10.18	1	10.58
3	1	10.23	1	11.08
4	1	10.36	1	11.81

Example 7.1. We show numerical examples of the optimum checkpoint intervals for a double modular system and a triple modular system when $\lambda S = 10^{-1}$. Table 7.1 presents $\lambda \widetilde{T}_1$ in (7.8), the optimum number N_1^*, λT_1^* and $\lambda L_1(N_1^*)$ for $\lambda C_1 = 0.5, 1, 1.5, 2, 3, 4, 5, 10, 20,$ and 30 ($\times 10^{-3}$). For example, when $\lambda = 10^{-2}(1/\text{sec})$, $C_1 = 10^{-1}(\text{sec})$, and $S = 10.0(\text{sec})$, the optimum number is $N_1^* = 5$, the optimum interval is $T_1^* = S/N^* = 2.0$ (sec), and the resulting mean time is $L_1(5) = 10.93$ (sec), that is about 9.3% longer than a native execution time S. In this case, note that $L_1(1) = 12.34$ (sec), *i.e.*, the mean time is about 88.6% shorter, compared with a noncheckpoint case.

Table 7.2 presents $\lambda \widetilde{T}_2$ in (7.13), N_2^*, λT_2^*, and $\lambda L_2(N_2^*)$ for a triple modular system. For example, when $\lambda = 10^{-2}$ (1/sec), $C_2 = 10^{-1}$ (sec), and $S = 10.0$ (sec), $N_2^* = 2$, $T_2^* = 5.0$ (sec), and $L_2(2) = 10.27$ (sec), that is about 2.7% longer than S. It can be easily seen in both tables that the more overheads C_i ($i = 1, 2$) increase, the more optimum numbers N_i^* decrease. The mean times of Table 7.1 are larger than those of Table 7.2 when $C_1 = C_2$.

Next, consider the problems of what majority system is optimum. When the overhead for the comparison of two states is C_1, it is assumed that the overhead for an $(n+1)$-out-of-$(2n+1)$ system is

$$C_{n+1} = \binom{2n+1}{2} C_1 \qquad (n = 1, 2, \dots),$$

because we select two states and compare them from each of $(2n+1)$ modules. Table 7.3 presents the optimum number N_{n+1}^* and the resulting mean time $\lambda L_{n+1}(N_{n+1}^*) \times 10^2$ of a majority decision system with $(2n+1)$ modules for $n = 1, 2, 3,$ and 4 when $\lambda C_1 = 0.1 \times 10^{-3}, 0.5 \times 10^{-3}$. When $\lambda C_1 = 0.5 \times 10^{-3}$, the optimum number is $N_3^* = 2$ and $\lambda L_2(2) = 10.37 \times 10^{-2}$, that is the smallest among these systems, *i.e.*, a 2-out-of-3 system is optimum. The mean times for $n = 1, 2$ when $\lambda C_1 = 0.5 \times 10^{-3}$ are smaller than 10.65×10^{-2} in Table 7.1 for a double modular system. ∎

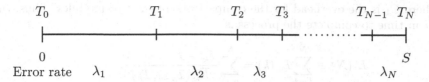

Fig. 7.5. Sequential checkpoint intervals and error rates

7.2 Sequential Checkpoint Intervals

It has been assumed in Sect. 7.1 that a native execution time S is divided equally, and an error rate λ is constant for any checkpoint intervals. In general, error rates would increase with the execution time, so that checkpoint intervals are not constant and should decrease with their numbers.

Suppose that checkpoints are placed at sequential times T_k ($k=1, 2, \ldots, N$), where $T_N \equiv S$ [147]. First, it is assumed that an error rate λ_k during an interval $(T_{k-1}, T_k]$ increases with k (Fig. 7.5). The mean time to complete the process successfully is obtained, and optimum checkpoint times that minimize it are derived numerically, using the similar method in Chap. 4. Furthermore, approximate checkpoint times are given by setting the probability of occurrences of errors for all checkpoint intervals constant.

Second, it is assumed that an error rate during $(T_{k-1}, T_k]$ increases with the original execution time, irrespective of the number of retries. Optimum checkpoint times that minimize the mean time to complete the process are discussed, and their approximate times are shown. Numerical examples of optimum checkpoint times for a double modular system are presented. It is shown numerically that the approximate method is simple, and these times become good approximations to optimum ones.

In this section, we consider only a double modular system as a redundant system. Using the similar methods in **(2)** and **(3)** of Sect. 7.1 and modifying them, these results could be extended to a majority decision system.

(1) Increasing Error Rate

Suppose that S is a native execution time of the process for a double modular system in **(1)** of Sect. 7.1. We divide S into N unequal parts and place a checkpoint at sequential times T_k ($k = 1, 2, \ldots, N - 1$), where $T_0 \equiv 0$ and $T_N \equiv S$. It is assumed that an error rate of one module during $(T_{k-1}, T_k]$ is λ_k that increases with k, i.e., $\lambda_k \leq \lambda_{k+1}$. Then, the probability that two modules have no error during $(T_{k-1}, T_k]$ is $\overline{F}_k(T_{k-1}, T_k) \equiv \exp[-2\lambda_k(T_k - T_{k-1})]$. Thus, by a similar method for obtaining (7.2), the mean execution time during $(T_{k-1}, T_k]$ is

$$L_1(k) = \frac{T_k - T_{k-1} + C_1}{\overline{F}_k(T_{k-1}, T_k)}, \tag{7.16}$$

where C_1 is the overhead for the comparisons of the two modules. Thus, the mean time to complete the process is

$$L_1(N) \equiv \sum_{k=1}^{N} L_1(k) = \sum_{k=1}^{N} \frac{T_k - T_{k-1} + C_1}{\overline{F}_k(T_{k-1}, T_k)}$$

$$= \sum_{k=1}^{N} (T_k - T_{k-1} + C_1) \exp[2\lambda_k (T_k - T_{k-1})]. \qquad (7.17)$$

We find optimum times T_k^* that minimize $L_1(N)$. Differentiating $L_1(N)$ with respect to T_k and setting it equal to zero,

$$[1 + 2\lambda_k (T_k - T_{k-1} + C_1)] \exp[2\lambda_k (T_k - T_{k-1})]$$
$$= [1 + 2\lambda_{k+1}(T_{k+1} - T_k + C_1)] \exp[2\lambda_{k+1}(T_{k+1} - T_k)]. \qquad (7.18)$$

Setting $x_k \equiv T_k - T_{k-1}$, where x_k represents the checkpoint interval, and rewriting (7.18) as a function of x_k,

$$\frac{1 + 2\lambda_{k+1}(x_{k+1} + C_1)}{1 + 2\lambda_k(x_k + C_1)} - \exp[2(\lambda_k x_k - \lambda_{k+1} x_{k+1})] = 0$$

$$(k = 1, 2, \ldots, N - 1). \qquad (7.19)$$

It is easily noted that $\lambda_{k+1} x_{k+1} \leq \lambda_k x_k$, and hence, $x_{k+1} \leq x_k$ because $\lambda_k \leq \lambda_{k+1}$. In particular, when $\lambda_{k+1} = \lambda_k \equiv \lambda$, $x_{k+1} = x_k \equiv T$ that corresponds to the standard checkpoint model in Sect. 7.1.

If $\lambda_{k+1} > \lambda_k$, then $x_{k+1} < x_k$. Let $Q_1(x_{k+1})$ be the left-hand side of (7.19) for a fixed x_k. Then, $Q_1(x_{k+1})$ increases strictly from

$$Q_1(0) = \frac{1 + 2\lambda_{k+1} C_1}{1 + 2\lambda_k(x_k + C_1)} - \exp(2\lambda_k x_k)$$

to $Q_1(x_k) > 0$. Thus, if $Q_1(0) < 0$, then an optimum x_{k+1}^* $(0 < x_{k+1}^* < x_k)$ to satisfy (7.19) exists uniquely.

Noting that $T_0 = 0$ and $T_N = S$, we have the following result:

(i) When $N = 1$, $T_1 = S$ and the mean time $L_1(S)$ is given in (7.5).
(ii) When $N = 2$, (7.19) is simplified as

$$[1 + 2\lambda_1 (x_1 + C_1)]e^{2\lambda_1 x_1} - [1 + 2\lambda_2(S - x_1 + C_1)]e^{2\lambda_2(S - x_1)} = 0. \quad (7.20)$$

Letting $Q_2(x_1)$ be the left-hand of (7.20), it increases strictly with x_1 from $Q_2(0) < 0$ to

$$Q_2(S) = [1 + 2\lambda_1 (S + C_1)]e^{2\lambda_1 S} - (1 + 2\lambda_2 C_1).$$

Thus, if $Q_2(S) > 0$, then $x_1^* = T_1^*$ $(0 < T_1^* < S)$ to satisfy (7.20) exists uniquely, and if $Q_2(S) \leq 0$, then $x_1^* = T_1^* = S$.

(iii) When $N = 3$, we compute x_k^* $(k = 1, 2)$ that satisfy the simultaneous equations:

$$[1 + 2\lambda_1(x_1 + C_1)]\,e^{2\lambda_1 x_1} = [1 + 2\lambda_2(x_2 + C_1)]\,e^{2\lambda_2 x_2}, \tag{7.21}$$

$$[1 + 2\lambda_2(x_2 + C_1)]\,e^{2\lambda_2 x_2} = [1 + 2\lambda_3(S - x_1 - x_2)]\,e^{2\lambda_3(S - x_1 - x_2)}. \tag{7.22}$$

(iv) When $N = 4, 5, \ldots$, we compute x_k^* and $T_k = \sum_{j=1}^{k} x_j^*$ similarly.

It is very troublesome to solve simultaneous equations numerically. We consider the following approximate checkpoint times: It is assumed that the probability that two modules have no error during any interval $(T_{k-1}, T_k]$ is constant, i.e., $\overline{F}_k\,(T_{k-1}, T_k) \equiv q$ $(k = 1, 2, \ldots, N)$. In this case, the mean time to complete the process is, from (7.17),

$$\widetilde{L}_1\,(N) = \frac{S + NC_1}{q}. \tag{7.23}$$

For example, when $\overline{F}_k\,(T_{k-1}, T_k) = e^{-2\lambda_k(T_k - T_{k-1})}$,

$$e^{-2\lambda_k(T_k - T_{k-1})} = q \equiv e^{-\widetilde{q}},$$

i.e.,

$$T_k - T_{k-1} = \frac{\widetilde{q}}{2\lambda_k}.$$

Thus,

$$\sum_{k=1}^{N} (T_k - T_{k-1}) = T_N = S = \widetilde{q} \sum_{k=1}^{N} \frac{1}{2\lambda_k}, \tag{7.24}$$

and

$$\widetilde{L}_1\,(N) = e^{\widetilde{q}}\,(S + NC_1). \tag{7.25}$$

Therefore, we compute $\widetilde{q}$ from (7.24) and $\widetilde{L}_1\,(N)$ from (7.25) for a specified S and N. Comparing $\widetilde{L}_1\,(N)$ for $N = 1, 2, \ldots$, we obtain an optimum $\widetilde{N}$ that minimizes $\widetilde{L}_1(N)$ and $\widetilde{q} = S / \sum_{k=1}^{\widetilde{N}} [1/(2\lambda_k)]$. Finally, we may compute $\widetilde{T}_k = \widetilde{q} \sum_{j=1}^{k} (1/2\lambda_j)$ $(k = 1, 2, \ldots, \widetilde{N} - 1)$.

Example 7.2. We compute the optimum sequential checkpoint times T_k^* and their approximate times $\widetilde{T}_k$ for a double modular system. It is assumed that $\lambda_k = [1 + \alpha\,(k - 1)]\lambda$ $(k = 1, 2, \ldots)$, i.e., an error rate increases by $100\alpha\%$ of an original rate λ. Table 7.4 presents the sequential times λT_k and the resulting mean times $\lambda L_1(N)$ for $N = 1, 2, \ldots$, and 9 when $\alpha = 0.1$, $\lambda S = 10^{-1}$, and $\lambda C_1 = 10^{-3}$. In this case, the mean time is the smallest when $N = 5$, i.e., the optimum checkpoint number is $N^* = 5$, the optimum checkpoint times T_k^* $(k = 1, 2, 3, 4, 5)$ should be placed at 2.38, 4.53, 6.50, 8.32, and 10.00 (sec) for $\lambda = 10^{-2}$ (1/sec), and the mean time 11.009 (sec) is about 10% longer than the native execution time $S = 10$ (sec). Note that checkpoint intervals

Table 7.4. Checkpoint intervals λT_k and mean time $\lambda L_1(N)$ when $\lambda_k = [1 + 0.1\,(k-1)]\lambda$, $\lambda S = 10^{-1}$, and $\lambda C_1 = 10^{-3}$

N	1	2	3	4	5
$\lambda T_1 \times 10^2$	10.00	5.24	3.65	2.85	2.38
$\lambda T_2 \times 10^2$		10.00	6.97	5.44	4.53
$\lambda T_3 \times 10^2$			10.00	7.81	6.50
$\lambda T_4 \times 10^2$				10.00	8.32
$\lambda T_5 \times 10^2$					10.00
$\lambda L_1\,(N) \times 10^2$	12.33617	11.32655	11.07923	11.00950	11.00887

N	6	7	8	9
$\lambda T_1 \times 10^2$	2.05	1.83	1.65	1.52
$\lambda T_2 \times 10^2$	3.91	3.48	3.15	2.89
$\lambda T_3 \times 10^2$	5.62	4.99	4.52	4.15
$\lambda T_4 \times 10^2$	7.19	6.39	5.78	5.31
$\lambda T_5 \times 10^2$	8.65	7.68	6.95	6.38
$\lambda T_6 \times 10^2$	10.00	8.88	8.03	7.37
$\lambda T_7 \times 10^2$		10.00	9.05	8.31
$\lambda T_8 \times 10^2$			10.00	9.18
$\lambda T_9 \times 10^2$				10.00
$\lambda L_1\,(N) \times 10^2$	11.04228	11.09495	11.15960	11.23220

$x_k = T_k - T_{k-1}$ decrease with k because error rates increase with the number of checkpoints.

Table 7.5 presents $\widetilde{q}$ and $\lambda \widetilde{L}_1(N)$ in (7.25) for $N = 1, 2, \ldots$, and 9 under the same assumptions as those in Table 7.4. In this case, $\widetilde{N} = N^* = 5$, and $\lambda \widetilde{L}_1(5) = 11.00888$ is a little longer than that in Table 7.4. When $\widetilde{N} = 5$, approximate optimum checkpoint times are $\lambda \widetilde{T}_k \times 10^2 = 2.37, 4.52, 6.49, 8.31$, and 10.00 that are a little shorter than those in Table 7.4. Such computations are much easier than solving simultaneous equations. It would be sufficient to adopt approximate checkpoint times as optimum ones in practical fields. ∎

We can apply the above results to a majority decision system in **(3)** of Sect. 7.1, denoting that

$$\overline{F}_k(T_{k-1}, T_k) = \sum_{j=n+1}^{2n+1} \binom{2n+1}{j} \{\exp\left[-\lambda_k\,(T_k - T_{k-1})\right]\}^j$$

$$\times \{1 - \exp\left[-\lambda_k\,(T_k - T_{k-1})\right]\}^{2n+1-j}$$

$$(k = 1, 2, \ldots, N). \qquad (7.26)$$

Table 7.5. Mean time $\lambda \widetilde{L}_1(N)$ for $\widetilde{q}$ when $\lambda S = 10^{-1}$ and $\lambda C_1 = 10^{-3}$

N	$\widetilde{q}$	$\lambda \widetilde{L}_1(N) \times 10^2$
1	0.2000000	12.33617
2	0.1047619	11.32655
3	0.0729282	11.07923
4	0.0569532	11.00951
5	0.0473267	11.00888
6	0.0408780	11.04229
7	0.0362476	11.09496
8	0.0327555	11.15962
9	0.0300237	11.23222

(2) General Error Rate

It is assumed that the probability that the system has no error during the checkpoint interval $(T_{k-1}, T_k]$ is $\overline{F}(T_k)/\overline{F}(T_{k-1})$ for a general modular system, irrespective of a rollback operation, where $F(t) \equiv 1 - \overline{F}(t)$. Then, the mean execution time during $(T_{k-1}, T_k]$ is

$$L_2(k) = (T_k - T_{k-1} + C) \frac{\overline{F}(T_k)}{\overline{F}(T_{k-1})}$$

$$+ [T_k - T_{k-1} + C + L_2(k)] \frac{F(T_k) - F(T_{k-1})}{\overline{F}(T_{k-1})},$$

and solving it,

$$L_2(k) = \frac{(T_k - T_{k-1} + C)\overline{F}(T_{k-1})}{\overline{F}(T_k)}.$$

Thus, the mean time to complete the process is

$$L_2(N) = \sum_{k=1}^{N} \frac{(T_k - T_{k-1} + C)\overline{F}(T_{k-1})}{\overline{F}(T_k)}. \tag{7.27}$$

We find optimum times T_k that minimize $L_2(N)$ for a specified N. Let $f(t)$ be a density function of $F(t)$ and $h(t) \equiv f(t)/\overline{F}(t)$ be the failure rate of $F(t)$. Then, differentiating $L_2(N)$ with respect to T_k and setting it equal to zero,

$$\frac{\overline{F}(T_{k-1})}{\overline{F}(T_k)}[1 + h(T_k)(T_k - T_{k-1} + C)] = \frac{\overline{F}(T_k)}{\overline{F}(T_{k+1})}[1 + h(T_k)(T_{k+1} - T_k + C)]$$

$$(k = 1, 2, \ldots, N - 1). \tag{7.28}$$

Therefore, we have the following result:

(i) When $N = 1$, $T_1 = S$ and the mean time is

$$L_2(1) = \frac{S + C}{\overline{F}(s)}. \tag{7.29}$$

(ii) When $N = 2$, (7.28) is

$$\frac{1}{\overline{F}(T_1)}[1 + h(T_1)(T_1 + C)] - \frac{\overline{F}(T_1)}{\overline{F}(S)}[1 + h(T_1)(S - T_1 + C)] = 0. \tag{7.30}$$

Letting $Q_2(T_1)$ be the left-hand side of (7.30), it is clearly seen that

$$Q_2(0) = 1 + h(0)C - \frac{1}{\overline{F}(S)}[1 + h(0)(S + C)] < 0,$$

$$Q_2(S) = \frac{1}{\overline{F}(S)}[1 + h(S)(S + C)] - [1 + h(S)C] > 0.$$

Thus, there exists a finite T_1 $(0 < T_1 < S)$ that satisfies (7.30).

(iii) When $N = 3$, we compute T_k $(k = 1, 2)$ that satisfy the simultaneous equations:

$$\frac{1}{\overline{F}(T_1)}[1 + h(T_1)(T_1 + C)] = \frac{\overline{F}(T_1)}{\overline{F}(T_2)}[1 + h(T_1)(T_2 - T_1 + C)], \tag{7.31}$$

$$\frac{\overline{F}(T_1)}{\overline{F}(T_2)}[1 + h(T_2)(T_2 - T_1 + C)] = \frac{\overline{F}(T_2)}{\overline{F}(S)}[1 + h(T_2)(S - T_2 + C)]. \tag{7.32}$$

(iv) When $N = 4, 5, \ldots$, we compute T_k similarly.

Next, we consider approximate times similar to those of (1). It is assumed that the probability that the system has no error during any interval (T_{k-1}, T_k) is constant, i.e., $\overline{F}(T_k)/\overline{F}(T_{k-1}) = q$. In this case, the mean time to complete the process is given in (7.23).

For example, when $\overline{F}(t) = \exp[-2(\lambda t)^m]$ for a double modular system,

$$\frac{\overline{F}(T_k)}{\overline{F}(T_{k-1})} = \exp\{-2[(\lambda T_k)^m - (\lambda T_{k-1})^m]\} = q \equiv e^{-\tilde{q}},$$

i.e.,

$$2(\lambda T_k)^m - 2(\lambda T_{k-1})^m = \tilde{q}.$$

Thus,

$$2(\lambda T_k)^m = k\tilde{q} \qquad (k = 1, 2, \ldots, N - 1),$$

$$2(\lambda T_N)^m = 2(\lambda S)^m = N\tilde{q}. \tag{7.33}$$

The mean time $\tilde{L}_2(N)$ to complete the process is given in (7.25).

Table 7.6. Checkpoint intervals λT_k and $\lambda L_2(N)$ when $\lambda C = 10^{-3}$ and $\lambda S = 10^{-1}$

N	1	2	3	4	5
$\lambda T_1 \times 10^2$	10.00	5.17	3.51	2.67	2.16
$\lambda T_2 \times 10^2$		10.00	6.80	5.17	4.18
$\lambda T_3 \times 10^2$			10.00	7.60	6.15
$\lambda T_4 \times 10^2$				10.00	8.09
$\lambda T_5 \times 10^2$					10.00
$\lambda L_2(N) \times 10^2$	11.83902	11.04236	10.85934	10.82069	10.83840

N	6	7	8	9
$\lambda T_1 \times 10^2$	1.81	1.57	1.38	1.23
$\lambda T_2 \times 10^2$	3.51	3.03	2.67	2.39
$\lambda T_3 \times 10^2$	5.17	4.46	3.93	3.51
$\lambda T_4 \times 10^2$	6.80	5.87	5.17	4.62
$\lambda T_5 \times 10^2$	8.41	7.26	6.39	5.71
$\lambda T_6 \times 10^2$	10.00	8.63	7.60	6.80
$\lambda T_7 \times 10^2$		10.00	8.81	7.87
$\lambda T_8 \times 10^2$			10.00	8.94
$\lambda T_9 \times 10^2$				10.00
$\lambda L_2(N) \times 10^2$	10.88391	10.94517	11.01622	11.09376

Example 7.3.   We compute the optimum sequential checkpoint times T_k^* and their approximate times $\widetilde{T}_k$ when error rates increase with the original execution time. It is assumed that $\overline{F}(t) = \exp[-2(\lambda t)^m]$ $(m > 1)$, $\lambda C = 10^{-3}$, and $\lambda S = 10^{-1}$ for a double modular system. Table 7.6 presents the sequential times λT_k and the resulting mean times $\lambda L_2(N)$ for $N = 1,\ 2,\ \ldots,$ and 9 when $\overline{F}(t) = \exp[-2(\lambda t)^{1.1}]$. In this case, the mean time is the smallest when $N = 4$, i.e., the optimum checkpoint number is $N^* = 4$, the checkpoint times T_k^* $(k = 1, 2, 3, 4)$ should be placed at 2.67, 5.17, 7.60, and 10.00 (sec) for $\lambda = 10^{-2}(1/\text{sec})$, and the mean time 10.8207 (sec) is about 8.2% longer than $S = 10$ (sec).

Table 7.7 presents $\widetilde{q}$ and $\lambda \widetilde{L}_2(N)$ for $N = 1,\ 2,\ \ldots,$ and 9 under the same assumptions as those in Table 7.6. In this case, $\widetilde{N} = N^* = 4$ and approximate checkpoint times are $\lambda \widetilde{T}_k \times 10^2 = 2.84,\ 5.33,\ 7.70,$ and 10.00, that are a little longer than those of Table 7.6, however, the mean time $\widetilde{L}_2(4)$ is almost the same in Table 7.6. ∎

For a majority decision system, we may denote that

Table 7.7. Mean time $\lambda\widetilde{L}_2(N)$ for $\widetilde{q}$ when $\lambda S = 10^{-1}$ and $\lambda C = 10^{-3}$

N	$\widetilde{q}$	$\lambda\widetilde{L}_2(N) \times 10^2$
1	0.1588656	11.83902
2	0.0794328	11.04326
3	0.0529552	10.86014
4	0.0397164	10.82136
5	0.0317731	10.83897
6	0.0264776	10.88441
7	0.0226951	10.94561
8	0.0198582	11.01661
9	0.0176517	11.09411

$$\frac{\overline{F}(T_k)}{\overline{F}(T_{k-1})} \equiv \sum_{j=n+1}^{2n+1} \binom{2n+1}{j} \left[\frac{\overline{F}_1(T_k)}{\overline{F}_1(T_{k-1})}\right]^j \left[\frac{F_1(T_k) - F_1(T_{k-1})}{\overline{F}_1(T_{k-1})}\right]^{2n+1-j}$$

$$(k = 1, 2, \ldots, N), \qquad (7.34)$$

where $\overline{F}_1(T_k)/\overline{F}_1(T_{k-1})$ is the probability that one module has no error during $(T_{k-1}, T_k]$.

7.3 Modified Checkpoint Models

When permanent failures in a double modular system have occurred, it would be impossible to detect them by comparing two states of each module. When two states have not matched at checkpoint times, we prepare another spare module for reexecuting the process of this interval [148]. The mean time to complete the process of one task is obtained, and optimum checkpoint times are computed numerically and compared with those of the standard modular system with no spare.

Next, we adopt three types of checkpoints for a double modular system: store-checkpoint (SCP), compare-checkpoint (CCP), and compare-and-store-checkpoint (CSCP) that combines SCP and CCP and has the same function as that of the checkpoint in the standard checkpoint model with overhead C_1 [149]. The mean times to the process of one task for the three checkpoint schemes are obtained, and optimum schemes are compared numerically with each other.

7.3.1 Double Modular System with Spare Process

Consider the same double modular system in (1) of Sect. 7.1. We call the process for the interval $((k-1)T, kT]$ that of task I_k $(k = 1, 2, \ldots, N)$. If two

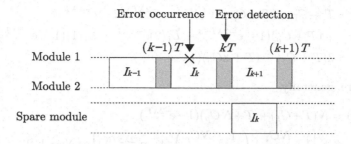

Fig. 7.6. Recovery scheme with a spare module

states of modules for task I_k match, two modules are correct and go forward to the process of task I_{k+1}. However, if two states do not match, a spare module makes the process of task I_k, and two modules make the process of task I_{k+1} (Fig. 7.6). It is assumed that the spare module has no error. If two states for task I_N match or the spare module makes the process of task I_N, the process completes one task successfully.

Let C_1 be the overhead for the comparison of two states and C_p be the total overhead for preparing a spare module and setting the correct process at checkpoint times, where $C_p \geq C_1$. Then, we compute the mean time $L_1(N)$ to complete the process of one task successfully.

In particular, when $N = 1$,

$$L_1(1) = (T + C_1)e^{-2\lambda T} + (T + C_1 + T + C_p)(1 - e^{-2\lambda T})$$
$$= T + C_1 + (T + C_p)(1 - e^{-2\lambda T}). \tag{7.35}$$

Furthermore, when $N = 2$ and $N = 3$, respectively,

$$L_1(2) = (T + C_1 + L_1(1))e^{-2\lambda T}$$
$$+ (T + C_1 + T + C_p + C_1)(1 - e^{-2\lambda T})e^{-2\lambda T}$$
$$+ (T + C_1 + T + C_p + C_1 + T + C_p)(1 - e^{-2\lambda T})^2$$
$$= T + C_1 + (T + C_p + C_1)(1 - e^{-2\lambda T})$$
$$+ (T + C_p)(1 - e^{-2\lambda T})^2 + L_1(1)e^{-2\lambda T}$$
$$= 2(T + C_1) + (T + 2C_p)(1 - e^{-2\lambda T}), \tag{7.36}$$

$$L_1(3) = [T + C_1 + L_1(2)]e^{-2\lambda T}$$
$$+ [T + C_1 + T + C_p + C_1 + L_1(1)](1 - e^{-2\lambda T})e^{-2\lambda T}$$
$$+ [T + C_1 + 2(T + C_p + C_1)](1 - e^{-2\lambda T})^2 e^{-2\lambda T}$$
$$+ [T + C_1 + 2(T + C_p + C_1) + T + C_p](1 - e^{-2\lambda T})^3$$

$$= T + C_1 + (T + C_p + C_1)(1 - e^{-2\lambda T})(2 - e^{-2\lambda T})$$
$$+ (T + C_p)(1 - e^{-2\lambda T})^3 + L_1(2)e^{-2\lambda T} + L_1(1)(1 - e^{-2\lambda T})e^{-2\lambda T}$$
$$= 3(T + C_1) + (T + 3C_p)(1 - e^{-2\lambda T}). \tag{7.37}$$

Therefore, generally,

$$L_1(N) = N(T + C_1) + (T + NC_p)(1 - e^{-2\lambda T})$$
$$= S\left[1 + \frac{NC_1}{S} + \left(\frac{1}{N} + \frac{NC_p}{S}\right)(1 - e^{-2\lambda S/N})\right] \quad (N = 1, 2, \dots). \tag{7.38}$$

We find an optimum number N^* that minimizes $L_1(N)$. Because $\lim_{N \to \infty} L_1(N) = \infty$, there exists a finite number N^* $(1 \le N^* < \infty)$. Setting $T \equiv S/N$ in (7.38) and rewriting it in terms of the function of T,

$$L_1(T) = S + \frac{SC_1}{T} + \left(T + \frac{SC_p}{T}\right)(1 - e^{-2\lambda T}). \tag{7.39}$$

Because $\lim_{T \to 0} L_1(T) = \lim_{T \to \infty} L_1(T) = \infty$, there exists an optimum $\widetilde{T}_1$ $(0 < \widetilde{T}_1 < \infty)$ that minimizes $L_1(T)$ in (7.39). Differentiating $L_1(T)$ with respect to T and setting it equal to zero,

$$T^2(2\lambda T e^{-2\lambda T} + 1 - e^{-2\lambda T}) = SC_1 + TC_p[1 - (1 + 2\lambda T)e^{-2\lambda T}]. \tag{7.40}$$

From the assumption $C_p \ge C_1$,

$$SC_1 \le \frac{T^2\left(2\lambda T e^{-2\lambda T} + 1 - e^{-2\lambda T}\right)}{2 - (1 + 2\lambda T)e^{-2\lambda T}} \le SC_p. \tag{7.41}$$

In addition, letting

$$Q(T) \equiv \frac{T^2(2\lambda T e^{-2\lambda T} + 1 - e^{-2\lambda T})}{2 - (1 + 2\lambda T)e^{-2\lambda T}},$$

it is easily noted that $Q(T)$ increases strictly from 0 to ∞. Thus, denoting T_c and T_p by solutions of equations $Q(T) = SC_1$ and $Q(T) = SC_p$, respectively, $T_c \le \widetilde{T}_1 \le T_p$. Therefore, using the partition method in Sect. 3.1, we get an optimum checkpoint number N_1^*.

In the particular case of $C_1 = C_p$, $\widetilde{T}_1$ is given by a finite and unique solution of $Q(T) = SC_1$. It can be clearly seen that $\widetilde{T}_1$ becomes longer, i.e., N_1^* becomes shorter, as the overhead C_1 becomes larger. Furthermore, using the approximation of $e^{-a} \approx 1 - a$ for small a, the mean time $L_1(T)$ in (7.39) is simplified as

$$\widetilde{L}_1(T) = S + \frac{SC_1}{T} + 2\lambda T^2 + 2\lambda SC_p. \tag{7.42}$$

Thus, an approximation time $\widetilde{T}_1$ that minimizes $\widetilde{L}_1(T)$ is

Table 7.8. Optimum time λT_1^*, number N_1^*, its approximate time $\lambda \widetilde{T}_1$, and mean time $\lambda L_1(N_1^*)$

	$\lambda C_1 \times 10^3$								
	1.0			2.0			10.0		
$\lambda C_p \times 10^2$	λT_1^* $\times 10^2$	N_1^*	$\lambda L_1(N_1^*)$ $\times 10^2$	λT_1^* $\times 10^2$	N_1^*	$\lambda L_1(N_1^*)$ $\times 10^2$	λT_1^* $\times 10^2$	N_1^*	$\lambda L_1(N_1^*)$ $\times 10^2$
0.1	2.97	3	10.53	-	-	-	-	-	-
0.5	2.98	3	10.61	3.76	3	10.91	-	-	-
1.0	2.98	3	10.71	3.77	3	11.01	6.52	2	12.67
2.0	3.00	3	10.90	3.79	3	11.20	6.54	2	12.86
3.0	3.02	3	11.10	3.81	3	11.40	6.56	2	13.05
5.0	3.06	3	11.48	3.84	3	11.78	6.59	2	13.43
10.0	3.15	3	12.45	3.93	3	12.75	6.68	2	14.38
50.0	4.10	3	20.19	4.83	2	20.39	7.50	1	21.88
100.0	5.85	2	29.71	6.40	2	29.91	8.77	1	30.94
200.0	10.43	1	48.17	10.68	1	48.27	12.20	1	49.07
$\lambda \widetilde{T}_1 \times 10^2$	2.92			3.68			6.30		

$$\widetilde{T}_1 = \left(\frac{S C_1}{4 \lambda} \right)^{1/3} . \qquad (7.43)$$

Example 7.4. Table 7.8 presents the optimum time λT_1^*, number N_1^*, and $\lambda L_1(N_1^*)$ for $\lambda C_1 = 1.0, 2.0,$ and $10.0 \ (\times 10^{-3})$ and $\lambda C_p = 0.1, 0.5, 1.0, 2.0, 3.0, 5.0, 10.0, 50.0, 100.0,$ and $200.0 \ (\times 10^{-2})$. When $\lambda = 10^{-2}$ (1/sec), $C_1 = 10^{-1}$ (sec), $C_p = 1.0$ (sec), and $S = 10.0$ (sec), the optimum number is $N_1^* = 3$ and the resulting mean time is $L_1(3) = 10.71$ (sec), that is about 4% shorter than $L_1(1) = 11.15$ (sec). This indicates that the optimum numbers N_1^* decrease slowly with C_p, and the approximate times $\widetilde{T}_1$ in (7.43) give a good lower bound of optimum times T_1^* for a small C_p.

Furthermore, comparing with Table 7.1 in (1) of Sect. 7.1 for a double modular system with no spare and Table 7.8 when $\lambda C_1 = 10^{-3}$, if λC_p is less than 2.0×10^{-2}, then a spare module should be provided, and conversely, if λC_p is larger than 3.0×10^{-2}, then it should not be done. ∎

7.3.2 Three Detection Schemes

We adopt three types of checkpoints CSCP, SCP, and CCP for a double modular system with constant error rate λ in (1) of Sect. 7.1: Suppose that

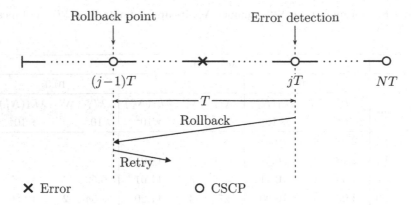

Fig. 7.7. Task execution for Scheme 1

S is a native execution time, and a CSCP is placed at periodic times kT $(k = 1, 2, \ldots, N)$, where $T \equiv S/N$, and either CCP or SCP is placed between CSCPs. To detect errors, we execute two modules and compare two states at time kT. Then, we introduce the following three overheads: C_s is the overhead to store the states of two modules, C_1 is the overhead to compare two states, and C_r is the overhead to roll back two modules to the previous checkpoint. Under the above assumptions, we consider three schemes that combine three types of checkpoints and obtain the mean time to complete the process of one task successfully.

(1) Scheme 1

Consider the same model with three overheads C_s, C_1, and C_r in **(1)** of Sect. 7.1, where two overheads are needed at every time kT and C_r is needed when some errors have occurred during $((k-1)T, kT]$ (Fig. 7.7). Then, using the method similar to obtaining (7.1),

$$L_1(1) = (T + C_s + C_1)\,e^{-2\lambda T} + [T + C_s + C_1 + C_r + L_1\,(1)]\,(1 - e^{-2\lambda T}),$$
$$(7.44)$$

and solving it,

$$L_1(1) = (T + C_s + C_1)\,e^{2\lambda T} + C_r(e^{2\lambda T} - 1). \qquad (7.45)$$

Thus, the mean time to complete the process is

$$L_1(N) \equiv NL_1\,(1) = N[(T + C_s + C_1)\,e^{2\lambda T} + C_r(e^{2\lambda T} - 1)]. \qquad (7.46)$$

Setting $T = S/N$ in (7.46),

$$L_1\,(N) = [S + N\,(C_s + C_1 + C_r)]e^{2\lambda S/N} - NC_r, \qquad (7.47)$$

or

$$L_1(T) = S\left[\left(1 + \frac{C_s + C_1 + C_r}{T}\right)e^{2\lambda T} - \frac{C_r}{T}\right]. \qquad (7.48)$$

When $C_s = C_r = 0$, $L_1(N)$ and $L_1(T)$ agree with (7.4) and (7.6), respectively.

Note that $\lim_{N\to\infty} L_1(N) = \lim_{T\to 0} L_1(T) = \infty$. Differentiating $L_1(T)$ in (7.48) with respect to T and setting it equal to zero,

$$2\lambda T(T + C_s + C_1 + C_r) - C_r(1 - e^{-2\lambda T}) = C_s + C_1. \qquad (7.49)$$

It is easily proved that the left-hand side of (7.49) increases strictly from 0 to ∞. Thus, if

$$2\lambda S(S + C_s + C_1 + C_r) - C_r(1 - e^{-2\lambda S}) \geq C_s + C_1, \qquad (7.50)$$

then there exists a finite and unique $\tilde{T}_1$ $(0 < \tilde{T}_1 \leq S)$ that satisfies (7.49).

Clearly,

$$2\lambda S(S + C_s + C_1 + C_r) - C_r(1 - e^{-2\lambda S}) > 2\lambda S(S + C_s + C_1).$$

If

$$2\lambda S(S + C_s + C_1) > C_s + C_1,$$

i.e.,

$$S > \frac{1}{2}\left[-(C_s + C_1) + \sqrt{(C_s + C_1)^2 + \frac{2(C_s + C_1)}{\lambda}}\right], \qquad (7.51)$$

then $0 < \tilde{T}_1 < S$. In addition, because

$$\frac{1}{2}\left[-(C_s + C_1) + \sqrt{(C_s + C_1)^2 + \frac{2(C_s + C_1)}{\lambda}}\right] < \sqrt{\frac{C_s + C_1}{2\lambda}},$$

if $S > \sqrt{(C_s + C_1)/2\lambda}$, then $0 < \tilde{T}_1 < S$.

From the above discussions, we can obtain an optimum number N_1^* that minimizes $L_1(N)$ in (7.47), using the partition method.

(2) Scheme 2

Suppose that a CSCP interval T is divided equally into m intervals, i.e., $m \equiv T/T_2$ (Fig. 7.8). A SCP is placed at time jT_2 $(j = 1, 2, \ldots, m-1)$ and the states of two modules are stored. If two states do not match at time kT, two modules are rolled back from kT to time $(j-1)T_2$ when some errors have occurred during $((j-1)T_2, jT_2]$ and reexecuted from $(j-1)T_2$.

The mean time for each interval $((k-1)T, kT]$ is

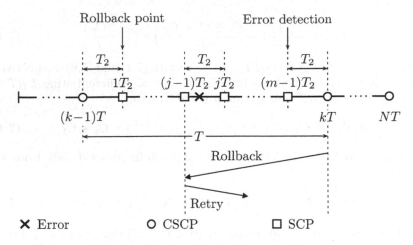

Fig. 7.8. Task execution for Scheme 2

$$L_2\left(m\right) = \left(mT_2 + mC_s + C_1\right) e^{-2\lambda mT_2}$$

$$+ \sum_{j=1}^{m} \int_{(j-1)T_2}^{jT_2} [mT_2 + mC_s + C_1 + C_r + L_2\left(m - (j-1)\right)]2\lambda e^{-2\lambda t}\, \mathrm{d}t$$

$$= mT_2 + mC_s + C_1 + C_r(1 - e^{-2\lambda mT_2})$$

$$+ \sum_{j=1}^{m} L_2\left(m - (j-1)\right)[e^{-2\lambda(j-1)T_2} - e^{-2\lambda jT_2}]. \tag{7.52}$$

Solving (7.52) for $L_2(m)$,

$$L_2(m) = \left(mT_2 + mC_s + C_1\right) e^{2\lambda T_2}$$

$$+ \left[\sum_{j=1}^{m-1} (jT_2 + jC_s + C_1) + mC_r\right](e^{2\lambda T_2} - 1)$$

$$= \left[(T_2 + C_s)\frac{m(m+1)}{2} + m(C_1 + C_r)\right](e^{2\lambda T_2} - 1)$$

$$+ mT_2 + mC_s + C_1, \tag{7.53}$$

that is equal to (7.45) when $m = 1$. Setting $T_2 = T/m$,

$$L_2(m) = \left[(T + mC_s)\frac{m+1}{2} + m(C_1 + C_r)\right](e^{2\lambda T/m} - 1)$$

$$+ T + mC_s + C_1 \qquad (m = 1, 2, \dots). \tag{7.54}$$

Because $\lim_{m \to \infty} L_2(m) = \infty$, there exists a finite m^* ($1 \le m^* < \infty$) that minimizes $L_2(m)$ in (7.54).

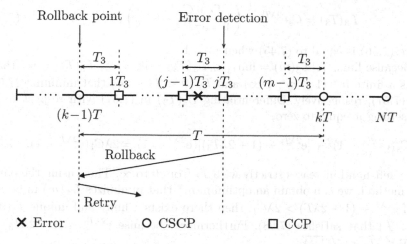

Fig. 7.9. Task execution for Scheme 3

(3) Scheme 3

Suppose that a CSCP interval T is also divided equally into m intervals, *i.e.*, $m \equiv T/T_3$ (Fig. 7.9). A CCP is placed at time jT_3 ($j = 1, 2, \ldots, m - 1$), and the states of two modules are compared at jT_3. When two states do not match at jT_3, some errors have occurred during $((j - 1)T_3, jT_3]$, and two modules are rolled back to time $(k - 1)T$ at which CSCP is placed.

The mean time for each interval $((k - 1)T, kT]$ is

$$L_3(m) = (mT_3 + mC_1 + C_s)\,\mathrm{e}^{-2\lambda mT_3}$$

$$+ \sum_{j=1}^{m-1} \int_{(j-1)T_3}^{jT_3} [jT_3 + jC_1 + C_r + L_3(m)]2\lambda\mathrm{e}^{-2\lambda t}\,\mathrm{d}t$$

$$+ \int_{(m-1)T_3}^{mT_3} [mT_3 + mC_1 + C_s + C_r + L_3(m)]2\lambda\mathrm{e}^{-2\lambda t}\,\mathrm{d}t,$$

and solving it,

$$L_3(m) = C_s\mathrm{e}^{2\lambda T_3} + \left(\frac{T_3 + C_1}{1 - \mathrm{e}^{-2\lambda T_3}} + C_r \right) (\mathrm{e}^{2\lambda mT_3} - 1). \qquad (7.55)$$

Setting $T_3 = T/m$,

$$L_3(m) = C_s\mathrm{e}^{2\lambda T/m} + \left(\frac{T/m + C_1}{1 - \mathrm{e}^{-2\lambda T/m}} + C_r \right) (\mathrm{e}^{2\lambda T} - 1) \quad (m = 1, 2, \ldots),$$

$$(7.56)$$

or

$$L_3(T_3) = C_s e^{2\lambda T_3} + \left(\frac{T_3 + C_1}{1 - e^{-2\lambda T_3}} + C_r\right)(e^{2\lambda T} - 1), \qquad (7.57)$$

where (7.56) is equal to (7.45) when $m = 1$.

Because $\lim_{m \to \infty} L_3(m) = \lim_{T_3 \to 0} L_3(T_3) = \lim_{T_3 \to \infty} L_3(T_3) = \infty$, there exists a finite m^* $(1 \le m^* < \infty)$ and $\widetilde{T}_3$ $(0 < \widetilde{T}_3 < \infty)$ that minimize (7.56) and (7.57), respectively. Differentiating $L_3(T_3)$ in (7.57) with respect to T_3 and setting it equal to zero,

$$2\lambda C_s (e^{2\lambda T_3} - 1)^2 + [e^{2\lambda T_3} - (1 + 2\lambda T_3)](e^{2\lambda T} - 1) = 2\lambda C_1 (e^{2\lambda T} - 1), \quad (7.58)$$

whose left-hand increases strictly with T_3 from 0 to ∞. Thus, using the partition method, we can obtain an optimum m^* that minimizes $L_3(m)$ in (7.56).

If $e^{2\lambda T} - (1 + 2\lambda T) > 2\lambda C_1$, then there exists a finite and unique $\widetilde{T}_3 (0 < \widetilde{T}_3 < T)$ that satisfies (7.58). Furthermore, because $e^{2\lambda T_3} - (1 + 2\lambda T_3) > 2(\lambda T_3)^2$, $\widetilde{T}_3 < \sqrt{C_1/\lambda}$.

Example 7.5. Suppose that $S = 1$, $C_1 = 2.5 \times 10^{-5}$, $C_s = 5 \times 10^{-4}$, and $C_r = 5 \times 10^{-4}$ [149]. Then, we present numerical examples of the optimum numbers of checkpoints and compare them among the three schemes.

Table 7.9 for Scheme 1 indicates $\widetilde{T}_1$ in (7.49), the optimum number N_1^*, the optimum time $T_1^* = 1/N_1^*$, the resulting mean time $L_1(N_1^*)$ in (7.47), and the mean time $L_1(1)$ for the no checkpoint case when $\lambda = 0.005$–0.500. For example, when $\lambda = 0.100$ (1/sec), $N_1^* = 20$, $T_1^* = 1/20 = 0.05$ (sec), and $L_1(20) = 1.02076$ (sec) that is about 2% longer than the native execution time $S = 1$ and $(1.22215 - 1.02950)/1.22215 = 16.5\%$ shorter than $L_1(1)$.

Table 7.10 for Scheme 2 indicates the optimum SCP numbers m^* between CSCPs, the optimum SCP interval $T_2^* = T_1^*/m^*$, the mean execution time $L_2(m^*)$ for a CSCP interval, and the resulting mean execution time $N_1^* L_2(m^*)$ to complete the process with the optimum intervals given in Table 7.9. This indicates that $m^* = 1$ for any λ, *i.e.*, we should place no SCP between SCP between CSCP intervals, because the overhead C_s to store the states of processors is 20 times longer than C_1 to compare the state.

Similarly, Table 7.11 for Scheme 3 presents the optimum CCP number m^* between CSCPs, the optimum CCP interval $T_3^* = T_1^*/m^*$, the mean execution time $L_3(m^*)$ for a CSCP interval, the mean time $N_1^* L_3(m^*)$ to complete the process, and the upper bound $\sqrt{C_1/\lambda}$ for $\widetilde{T}_3$. For example, when $\lambda = 0.100(1/\text{sec})$, the mean execution time is $20 \times L_3(3) = 1.01834$ (sec), that is about 0.00242 seconds shorter than $L_1(20)$. Furthermore, the upper bounds $\sqrt{C_1/\lambda}$ are smaller than T_3^* except when $\lambda = 0.005$, however, they would be helpful to estimate an optimum number m^* roughly by calculating $T_1^*/\sqrt{C_1/\lambda}$. It is natural that the mean times $N_1^* L_3(m^*)$ are smaller than $L_1(N_1^*)$ in Table 7.9. ∎

Table 7.9. Optimum CSCP number N_1^*, CSCP interval T_1^*, and mean execution time $L_1(N^*)$ for Scheme 1

λ	$\widetilde{T}_1$	N_1^*	T_1^*	$L_1(N_1^*)$	$L_1(1)$
0.005	0.22887	4	0.25000	1.00461	1.01059
0.010	0.16176	6	0.16667	1.00651	1.02075
0.050	0.07219	14	0.07143	1.01462	1.10580
0.100	0.05097	20	0.05000	1.02076	1.22215
0.200	0.03597	28	0.03571	1.02950	1.49285
0.300	0.02932	34	0.02941	1.03627	1.82349
0.400	0.02535	39	0.02564	1.04203	2.22732
0.500	0.02265	44	0.02273	1.04712	2.72057

Table 7.10. Optimum SCP number m^*, CSCP interval T_2^*, and mean execution time $N_1^* L_2(m^*)$ for Scheme 2

λ	m^*	T_2^*	$L_2(m^*)$	$N_1^* L_2(m^*)$
0.005	1	0.25000	0.25115	1.00461
0.010	1	0.16667	0.16775	1.00651
0.050	1	0.07143	0.07247	1.01462
0.100	1	0.05000	0.05104	1.02076
0.200	1	0.03571	0.03677	1.02950
0.300	1	0.02941	0.03048	1.03627
0.400	1	0.02564	0.02672	1.04203
0.500	1	0.02273	0.02380	1.04712

Table 7.11. Optimum CCP number m^*, CCP interval T_3^*, and mean execution time $N_1^* L_3(m^*)$ for Scheme 3

λ	m^*	T_3^*	$L_3(m^*)$	$N_1^* L_3(m^*)$	$\sqrt{C_1/\lambda}$
0.005	4	0.06250	0.25099	1.00397	0.07071
0.010	3	0.05556	0.16761	1.00569	0.05000
0.050	3	0.02381	0.07235	1.01290	0.02236
0.100	3	0.01667	0.05092	1.01834	0.01581
0.200	3	0.01190	0.03664	1.02597	0.01118
0.300	3	0.00980	0.03035	1.03183	0.00913
0.400	3	0.00855	0.02658	1.03679	0.00791
0.500	3	0.00758	0.02367	1.04131	0.00707

Maintenance Models with Two Variables

Almost all units deteriorate with age and use, and eventually, fail from either or both causes in a random environment. If their failure rates increase with age and use, it may be wise to do some maintenance when they reach a certain age or are used a certain number of times. This policy would be effective where units suffer great deterioration with both age and use. For example, some parts of aircraft have to be maintained at a specified number of flights and a planned time. This would be applied to the maintenance of some parts of large complex systems such as switching devices and parts of transportation equipment, computers, and plants.

A methodical survey of maintenance policies in reliability theory was done [1]. The recent published books [4, 72, 73, 150] collected many reliability and maintenance models and their optimum policies. Several maintenance policies for a finite interval are summarized in Chaps. 4 and 5. This chapter surveys widely used maintenance models with continuous and discrete variables and discusses analytically their optimum policies: Typical replacement policies are age, periodic, and block replacements. Sect. 8.1 takes up three replacement models with continuous time T and discrete number N, where the unit is replaced before failure at a planned time T or at some number N such as uses, working times, and failures, whichever occurs first. We investigate analytically properties of three replacement models and specify the computing procedure for obtaining both optimum T^* and N^* that minimize the expected cost rates.

It might be impossible to replace a working unit even when a planned time T comes and be wise to replace it at the first completion of the working time after time T. We sometimes want to use a working unit as long as possible, where the replacement cost after failure is not so high. From such viewpoints, we propose in Sect. 8.2 two modified models, where the unit is replaced at the Nth completion of working times over time T, and is replaced at time T or at the Nth working time, whichever occurs last. Furthermore, we consider the modified backup model with task N and backup time T that has been discussed in Chap. 7. Section. 8.3 takes up the replacement model of a parallel system with N units and replacement time T, the inspection models with

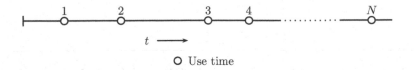

Fig. 8.1. Process of use times

periodic times, and the model of a storage system with replacement time NT. Optimum time T^* and number N^* that minimize the expected cost rates for each model are derived.

8.1 Three Replacement Models

Three policies of age, periodic, and block replacements have been well-known in reliability theory and commonly used in actual fields [1, 2]. This section summarizes a variety of three replacement models with continuous time T and discrete number N. Two models of age replacement are considered: (1) A unit is replaced before failure at a planned time T or at a number N of uses [1, p. 83, 151], and (2) it is replaced at a planned time T or at a number N of working times [152]. Three models of periodic replacement where the unit is replaced at a number of uses, failures, and type 1 failures are proposed [1, p. 110, 153–155]. Two models with random maintenance quality that are replaced at time NT are proposed [156]. In block replacement, the unit is replaced at time T or at the Nth failure. Expected cost rates for each model are obtained, and optimum time T^* and number N^* that minimize them are discussed analytically.

8.1.1 Age Replacement

Most systems deteriorate usually with both their operating time and number of uses. In such failure studies, the time to failure is observed mostly on operating time or calendar time. A good timescale of failure maintenance models was discussed [157]. For such systems, it may be wise to replace them at their total operating time and number of uses. On the other hand, when some systems in offices and industry successively execute jobs and computer processes, it would be better to replace them after they have completed their work and processes.

   This section takes up two age replacement policies: (1) A unit is replaced at time T or at number N of uses, and (2) it is replaced at time T or at number N of working times, whichever occurs first. The expected cost rates for two models are obtained, and both optimum T^* and N^* that minimize them are derived.

(1) Number of Uses

A unit begins to operate at time 0 and is used according to a renewal process with a general distribution $G(t)$ (Fig. 8.1) and its finite mean $1/\theta \equiv \int_0^\infty \overline{G}(t)\mathrm{d}t < \infty$, where, in general, $\overline{\Phi}(t) \equiv 1 - \Phi(t)$ for any function $\Phi(t)$. It is assumed that the usage time of the unit would be negligible because its time is very small compared with its age. The probability that the unit is used exactly j times during $[0, t]$ is $G^{(j)}(t) - G^{(j+1)}(t)$, where $G^{(j)}(t)$ $(j = 1, 2, \dots)$ denotes the j-fold Stieltjes convolution of $G(t)$ with itself and $G^{(0)}(t) \equiv 1$ for $t \geq 0$. The continuous distribution of failures due to deterioration with age is $F(t)$ with a finite mean μ_1, and the discrete distribution of failures due to use is $\{p_j\}_{j=1}^\infty$, where $F(t)$ and p_j are independent of each other. It is assumed that the failure rates of both distributions are $h(t) \equiv f(t)/\overline{F}(t)$ and $h_j \equiv p_j/(1 - P_{j-1})$, respectively, where $f(t)$ is a density function of $F(t)$, $P_j \equiv \sum_{i=1}^j p_i$ $(j = 1, 2, \dots)$, and $P_0 \equiv 0$.

Suppose that the unit is replaced before failure at a planned time T $(0 < T \leq \infty)$ of age or at a planned number N of uses, whichever occurs first. Then, the probability that the unit is replaced at time T is

$$\overline{F}(T) \sum_{j=0}^{N-1} (1 - P_j)[G^{(j)}(T) - G^{(j+1)}(T)], \tag{8.1}$$

the probability that it is replaced at number N is

$$(1 - P_N) \int_0^T \overline{F}(t)\,\mathrm{d}G^{(N)}(t), \tag{8.2}$$

and the probability that it is replaced at failure is

$$\sum_{j=0}^{N-1} \left\{ (1-P_j) \int_0^T [G^{(j)}(t) - G^{(j+1)}(t)]\,\mathrm{d}F(t) + p_{j+1} \int_0^T \overline{F}(t)\,\mathrm{d}G^{(j+1)}(t) \right\}, \tag{8.3}$$

where note that $(8.1) + (8.2) + (8.3) = 1$. The mean time to replacement is

$$T\overline{F}(T) \sum_{j=0}^{N-1} (1 - P_j)[G^{(j)}(T) - G^{(j+1)}(T)] + (1 - P_N) \int_0^T t\overline{F}(t)\,\mathrm{d}G^{(N)}(t)$$

$$+ \sum_{j=0}^{N-1} \left\{ (1 - P_j) \int_0^T t[G^{(j)}(t) - G^{(j+1)}(t)]\,\mathrm{d}F(t) + p_{j+1} \int_0^T t\overline{F}(t)\,\mathrm{d}G^{(j+1)}(t) \right\}$$

$$= \sum_{j=0}^{N-1} (1 - P_j) \int_0^T [G^{(j)}(t) - G^{(j+1)}(t)]\overline{F}(t)\,\mathrm{d}t. \tag{8.4}$$

Introduce the following costs: Cost c_F is the replacement cost at failure, and c_T and c_N are the respective replacement costs at time T and number N. Then, from (8.1)–(8.4), the expected cost rate [1, p. 71] is

$$C_1(T,N) = \frac{\begin{aligned}&c_F - (c_F - c_T)\overline{F}(T)\sum_{j=0}^{N-1}(1 - P_j)[G^{(j)}(T) - G^{(j+1)}(T)]\\&-(c_F - c_N)(1 - P_N)\int_0^T \overline{F}(t)dG^{(N)}(t)\end{aligned}}{\sum_{j=0}^{N-1}(1 - P_j)\int_0^T [G^{(j)}(t) - G^{(j+1)}(t)]\overline{F}(t)dt}.$$

(8.5)

This includes some basic replacement models: When the unit is replaced before failure only at time T,

$$\begin{aligned}C_1(T) &\equiv \lim_{N\to\infty} C_1(T,N)\\&= \frac{c_F - (c_F - c_T)\overline{F}(T)\sum_{j=1}^{\infty}p_j[1 - G^{(j)}(T)]}{\sum_{j=1}^{\infty}p_j\int_0^T [1 - G^{(j)}(t)]\overline{F}(t)dt},\end{aligned}$$

(8.6)

when it is replaced before failure only at number N,

$$\begin{aligned}C_1(N) &\equiv \lim_{T\to\infty} C_1(T,N)\\&= \frac{c_F - (c_F - c_N)(1 - P_N)\int_0^{\infty} G^{(N)}(t)dF(t)}{\sum_{j=0}^{N-1}(1 - P_j)\int_0^{\infty}[G^{(j)}(t) - G^{(j+1)}(t)]\overline{F}(t)dt},\end{aligned}$$

(8.7)

and when it is replaced only at failure,

$$C_1 \equiv \lim_{N\to\infty} C_1(N) = \lim_{T\to\infty} C_1(T) = \frac{c_F}{\sum_{j=1}^{\infty}p_j\int_0^{\infty}[1 - G^{(j)}(t)]\overline{F}(t)dt}.$$

(8.8)

The optimum policies that minimize $C_1(T)$ in (8.6) and $C_1(N)$ in (8.7) when $G(t) = 1 - e^{-\theta t}$, i.e., uses occur in a Poisson process with rate θ, were discussed analytically [1, p. 85, 151].

When $G(t) = 1 - e^{-\theta t}$ and $c_T = c_N < c_F$, the expected cost rate $C_1(T,N)$ in (8.5) is rewritten as

$$C_1(T,N) = \frac{\begin{aligned}&(c_F - c_T)\sum_{j=0}^{N-1}\Big\{(1 - P_j)\int_0^T [(\theta t)^j/j!]e^{-\theta t}dF(t)\\&+ p_{j+1}\int_0^T \theta[(\theta t)^j/j!]e^{-\theta t}\overline{F}(t)dt\Big\} + c_T\end{aligned}}{\sum_{j=0}^{N-1}(1 - P_j)\int_0^T [(\theta t)^j/j!]e^{-\theta t}\overline{F}(t)dt}.$$

(8.9)

We find both optimum T^* and N^* that minimize $C_1(T,N)$ when the failure rate $h(t)$ increases strictly to ∞ and h_j increases strictly. Differentiating $C_1(T,N)$ with respect to T and setting it equal to zero,

$$Q_1(T;N) = \frac{c_T}{c_F - c_T},$$

(8.10)

where

$Q_1(T; N)$

$$\equiv \left\{ h(T) + \frac{\theta \sum_{j=0}^{N-1} p_{j+1}[(\theta T)^j/j!]}{\sum_{j=0}^{N-1}(1 - P_j)[(\theta T)^j/j!]} \right\} \sum_{j=0}^{N-1}(1 - P_j) \int_0^T \frac{(\theta t)^j}{j!} e^{-\theta t} \overline{F}(t)\, dt$$

$$- \sum_{j=0}^{N-1} \left[(1 - P_j) \int_0^T \frac{(\theta t)^j}{j!} e^{-\theta t}\, dF(t) + p_{j+1} \int_0^T \frac{\theta(\theta t)^j}{j!} e^{-\theta t} \overline{F}(t)\, dt \right].$$

First, prove that when h_j increases strictly,

$$\frac{\sum_{j=0}^{N} p_{j+1}[(\theta T)^j/j!]}{\sum_{j=0}^{N}(1 - P_j)[(\theta T)^j/j!]} \tag{8.11}$$

also increases strictly with T and converges to h_{N+1} as $T \to \infty$ for any N [1, p. 85]. Differentiating (8.11) with respect to T,

$$\frac{\theta}{\left\{ \sum_{j=0}^{N}(1 - P_j)[(\theta T)^j/j!] \right\}^2} \left[\sum_{j=1}^{N} p_{j+1} \frac{(\theta T)^{j-1}}{(j-1)!} \sum_{j=0}^{N}(1 - P_j) \frac{(\theta T)^j}{j!} \right.$$

$$\left. - \sum_{j=0}^{N} p_{j+1} \frac{(\theta T)^j}{j!} \sum_{j=1}^{N}(1 - P_j) \frac{(\theta T)^{j-1}}{(j-1)!} \right].$$

The expression within the bracket of the numerator is

$$\sum_{j=1}^{N} \frac{(\theta T)^{j-1}}{(j-1)!} \sum_{i=0}^{N} \frac{(\theta T)^i}{i!}(1 - P_i)(1 - P_j)(h_{j+1} - h_{i+1})$$

$$= \sum_{j=1}^{N} \frac{(\theta T)^{j-1}}{(j-1)!} \sum_{i=0}^{j-1} \frac{(\theta T)^i}{i!}(1 - P_i)(1 - P_j)(h_{j+1} - h_{i+1})$$

$$+ \sum_{j=1}^{N} \frac{(\theta T)^{j-1}}{(j-1)!} \sum_{i=j}^{N} \frac{(\theta T)^i}{i!}(1 - P_i)(1 - P_j)(h_{j+1} - h_{i+1})$$

$$= \sum_{j=1}^{N} \frac{(\theta T)^{j-1}}{j!} \sum_{i=0}^{j-1} \frac{(\theta T)^i}{i!}(1 - P_i)(1 - P_j)(h_{j+1} - h_{i+1})(j - i) > 0,$$

that implies that (8.11) increases strictly with T. Furthermore, it can be easily proved that (8.11) tends to h_{N+1} as $T \to \infty$.

From the above results, $\lim_{T \to 0} Q_1(T; N) = 0$, $\lim_{T \to \infty} Q_1(T; N) = \infty$, and $Q_1(T; N)$ increases strictly with T. Thus, there exists a finite and unique T^* $(0 < T^* < \infty)$ that satisfies (8.10) for any $N \geq 1$, and the resulting cost rate is

$$C_1(T^*, N) = (c_F - c_T) \left\{ h(T^*) + \frac{\theta \sum_{j=0}^{N-1} p_{j+1}[(\theta T^*)^j/j!]}{\sum_{j=0}^{N-1}(1 - P_j)[(\theta T^*)^j/j!]} \right\}. \tag{8.12}$$

Next, from the inequality $C_1(T, N+1) - C_1(T, N) \geq 0$ for a fixed $T > 0$,

$$L_1(N; T) \geq \frac{c_T}{c_F - c_T} \qquad (N = 1, 2, \dots), \qquad (8.13)$$

where

$$L_1(N; T) \equiv \left[\theta h_{N+1} + \frac{\int_0^T (\theta t)^N e^{-\theta t} dF(t)}{\int_0^T (\theta t)^N e^{-\theta t} \overline{F}(t) dt}\right] \sum_{j=0}^{N-1} (1 - P_j) \int_0^T \frac{(\theta t)^j}{j!} e^{-\theta t} \overline{F}(t)\, dt$$

$$- \sum_{j=0}^{N-1} \left[(1 - P_j) \int_0^T \frac{(\theta t)^j}{j!} e^{-\theta t}\, dF(t) + p_{j+1} \int_0^T \frac{\theta(\theta t)^j}{j!} e^{-\theta t} \overline{F}(t)\, dt\right].$$

$$(8.14)$$

Second, prove that when $h(t)$ increases strictly,

$$\frac{\int_0^T (\theta t)^N e^{-\theta t} dF(t)}{\int_0^T (\theta t)^N e^{-\theta t} \overline{F}(t) dt} \qquad (8.15)$$

also increases strictly with N and converges to $h(T)$ as $N \to \infty$ for all $T > 0$ [1, p. 85]. Denoting

$$q(T) \equiv \int_0^T (\theta t)^{N+1} e^{-\theta t} dF(t) \int_0^T (\theta t)^N e^{-\theta t} \overline{F}(t) dt$$

$$- \int_0^T (\theta t)^N e^{-\theta t} dF(t) \int_0^T (\theta t)^{N+1} e^{-\theta t} \overline{F}(t) dt,$$

it follows that $\lim_{T \to 0} q(T) = 0$, and

$$\frac{dq(T)}{dT} = (\theta T)^N e^{-\theta T} \overline{F}(T) \int_0^T (\theta t)^N e^{-\theta t} \overline{F}(t)(\theta T - \theta t)[h(T) - h(t)]\, dt > 0.$$

Thus, $q(T)$ increases strictly with T from 0 for any $N \geq 0$, that implies that (8.15) increases strictly with N. Furthermore, from the assumption that $h(t)$ increases,

$$\frac{\int_0^T (\theta t)^N e^{-\theta t} dF(t)}{\int_0^T (\theta t)^N e^{-\theta t} \overline{F}(t) dt} \leq h(T).$$

On the other hand, for any $\delta \in (0, T)$,

$$\frac{\int_0^T (\theta t)^N e^{-\theta t} dF(t)}{\int_0^T (\theta t)^N e^{-\theta t} \overline{F}(t) dt} = \frac{\int_0^{T-\delta} (\theta t)^N e^{-\theta t} dF(t) + \int_{T-\delta}^T (\theta t)^N e^{-\theta t} dF(t)}{\int_0^{T-\delta} (\theta t)^N e^{-\theta t} \overline{F}(t) dt + \int_{T-\delta}^T (\theta t)^N e^{-\theta t} \overline{F}(t) dt}$$

$$\geq \frac{h(T - \delta) \int_{T-\delta}^T (\theta t)^N e^{-\theta t} \overline{F}(t) dt}{\int_0^{T-\delta} (\theta t)^N e^{-\theta t} \overline{F}(t) dt + \int_{T-\delta}^T (\theta t)^N e^{-\theta t} \overline{F}(t) dt}$$

$$= \frac{h(T - \delta)}{1 + \left[\int_0^{T-\delta} (\theta t)^N e^{-\theta t} \overline{F}(t) dt / \int_{T-\delta}^T (\theta t)^N e^{-\theta t} \overline{F}(t) dt\right]}.$$

The quantity in the bracket of the denominator is

$$\frac{\int_0^{T-\delta}(\theta t)^N e^{-\theta t}\overline{F}(t)dt}{\int_{T-\delta}^T(\theta t)^N e^{-\theta t}\overline{F}(t)dt} \le \frac{e^{\theta T}}{\delta\overline{F}(T)}\int_0^{T-\delta}\left(\frac{t}{T-\delta}\right)^N dt \to 0 \quad \text{as } N \to \infty.$$

Thus, it follows that

$$h(T-\delta) \le \lim_{N\to\infty}\frac{\int_0^T(\theta t)^N e^{-\theta t}dF(t)}{\int_0^T(\theta t)^N e^{-\theta t}\overline{F}(t)dt} \le h(T),$$

that completes the proof because δ is arbitrary.

From the above results,

$$L_1(N+1;T) - L_1(N;T) = \sum_{j=0}^N(1-P_j)\int_0^T\frac{(\theta t)^j}{j!}e^{-\theta t}\overline{F}(t)\,dt$$

$$\times\left[\theta(h_{N+2}-h_{N+1}) + \frac{\int_0^T(\theta t)^{N+1}e^{-\theta t}dF(t)}{\int_0^T(\theta t)^{N+1}e^{-\theta t}\overline{F}(t)dt} - \frac{\int_0^T(\theta t)^N e^{-\theta t}dF(t)}{\int_0^T(\theta t)^N e^{-\theta t}\overline{F}(t)dt}\right] > 0.$$

In addition, because T and N have to satisfy (8.10), the inequality (8.13) can be rewritten as

$$\theta\left\{h_{N+1} - \frac{\sum_{j=0}^{N-1}P_{j+1}[(\theta T)^j/j!]}{\sum_{j=0}^{N-1}(1-P_j)[(\theta T)^j/j!]}\right\} + \frac{\int_0^T(\theta t)^N e^{-\theta t}dF(t)}{\int_0^T(\theta t)^N e^{-\theta t}\overline{F}(t)dt} \ge h(T).$$

$$(8.16)$$

Note that the left-hand side of (8.16) is greater than $h(T)$ as $N \to \infty$. Hence, there exists a finite and unique minimum N^* ($1 \le N^* < \infty$) that satisfies (8.16).

From the above discussions, we can specify the computing procedure for obtaining optimum T^* and N^*:

(i) Compute a minimum N_1 to satisfy $L_1(N;\infty) \ge c_T/(c_F-c_T)$ from (8.13).
(ii) Compute T_k to satisfy (8.10) for $N_k(k=1,2,\ldots)$.
(iii) Compute a minimum N_{k+1} to satisfy (8.16) for $T_k(k=1,2,\ldots)$.
(iv) Continue the computation until $N_k = N_{k+1}$, and set $N_k = N^*$ and $T_k = T^*$.

Example 8.1. Suppose that $G(t) = 1 - e^{-t}$, $F(t)$ is a Weibull distribution $[1 - \exp(-\lambda t^2)]$, and $\{p_j\}$ is a negative binomial distribution $p_j = jp^2q^{j-1}$ ($j = 1,2,\ldots$), where $q \equiv 1-p$. In addition, assume that the mean time to failure $(1+q)/p$ caused by use is equal to $\sqrt{\pi}/(2\sqrt{\lambda})$ caused by deterioration with age, *i.e.*, $\lambda = \pi p^2/[4(1+q)^2]$. Table 8.1 presents the optimum T^* and N^*, and the expected cost rate $C_1(T^*,N^*)$ for p. Both T^* and N^* increase with p. It is of interest that T^* is a little longer than N^*. ∎

Table 8.1. Optimum T^*, N^* and expected cost rate $C_1(T^*, N^*)$

p	T^*	N^*	$C_1(T^*, N^*)$
0.1	6.1	5	0.5206
0.05	11.5	10	0.2457
0.02	26.8	25	0.0947
0.01	52.0	50	0.0469
0.005	101.6	99	0.0233

(2) Number of Working Times

A unit operates according to successive and independent working times with a general distribution $G(t)$ and its finite mean $1/\theta$. The probability that the unit operates exactly j working times during $(0, t]$ is $G^{(j)}(t) - G^{(j+1)}(t)$. The unit fails according to a general distribution $F(t)$ with a finite mean $\mu \equiv \int_0^\infty \overline{F}(t)dt < \infty$ and its failure rate $h(t)$, that is independent of the number of working times.

Suppose that the unit is replaced before failure at a planned time T or at the Nth completion of working times, whichever occurs first. Then, setting $p_j \equiv 0$ in (8.5), the expected cost rate is

$$C_2(T, N) = \frac{c_F - (c_F - c_T)\overline{F}(T)[1 - G^{(N)}(T)] - (c_F - c_N)\int_0^T \overline{F}(t)dG^{(N)}(t)}{\int_0^T [1 - G^{(N)}(t)]\overline{F}(t)dt}.$$

(8.17)

When the unit is replaced only at time T,

$$C_2(T) \equiv \lim_{N\to\infty} C_2(T, N) = \frac{c_F - (c_F - c_T)\overline{F}(T)}{\int_0^T \overline{F}(t)dt},$$

(8.18)

that agrees with (3.4) of [1, p. 72] for the standard age replacement, when it is replaced only at number N,

$$C_2(N) \equiv \lim_{T\to\infty} C_2(T, N) = \frac{c_F - (c_F - c_N)\int_0^\infty \overline{F}(t)dG^{(N)}(t)}{\int_0^\infty [1 - G^{(N)}(t)]\overline{F}(t)dt},$$

(8.19)

and when it is replaced only at failure,

$$C_2 \equiv \lim_{T\to\infty} C_2(T) = \lim_{N\to\infty} C_2(N) = \frac{c_F}{\mu}.$$

(8.20)

Furthermore, when $N = 1$, this corresponds to the age replacement model with a random working time according to a distribution $G(t)$ [158].

We find both optimum T^* and N^* that minimize $C_2(T, N)$ in (8.17) when $c_T = c_N < c_F$ and the failure rate $h(t)$ increases strictly to ∞. Differentiating $C_2(T, N)$ with respect to T and setting it equal to zero,

$$h(T) \int_0^T [1 - G^{(N)}(t)]\overline{F}(t)\, dt - \int_0^T [1 - G^{(N)}(t)]\, dF(t) = \frac{c_T}{c_F - c_T}. \quad (8.21)$$

It can be clearly seen that the left-hand side of (8.21) increases strictly with T from 0 to ∞. Thus, there exists a finite and unique T^* $(0 < T^* < \infty)$ that satisfies (8.21) for any $N \geq 1$, and the resulting cost rate is

$$C_2(T^*, N) = (c_F - c_T)h(T^*). \quad (8.22)$$

It is also proved that optimum T^* decreases with N because the left-hand side of (8.21) increases with N.

Next, from the inequality $C_2(T, N + 1) - C_2(T, N) \geq 0$,

$$\frac{\int_0^T [G^{(N)}(t) - G^{(N+1)}(t)]dF(t)}{\int_0^T [G^{(N)}(t) - G^{(N+1)}(t)]\overline{F}(t)dt} \int_0^T [1 - G^{(N)}(t)]\overline{F}(t)\, dt$$

$$- \int_0^T [1 - G^{(N)}(t)]\, dF(t) \geq \frac{c_T}{c_F - c_T}. \quad (8.23)$$

It is easily proved that if

$$\frac{\int_0^T [G^{(N)}(t) - G^{(N+1)}(t)]dF(t)}{\int_0^T [G^{(N)}(t) - G^{(N+1)}(t)]\overline{F}(t)dt} \quad (8.24)$$

increases strictly with N, the left-hand side of (8.23) also increases strictly with N for any $T > 0$. Thus, if a finite N exists such that (8.23) holds, an optimum N^* is derived by a unique minimum that satisfies (8.23). In addition, the inequality (8.23) is rewritten as, from (8.21),

$$\frac{\int_0^T [G^{(N)}(t) - G^{(N+1)}(t)]dF(t)}{\int_0^T [G^{(N)}(t) - G^{(N+1)}(t)]\overline{F}(t)dt} \geq h(T).$$

However, from the assumption that $h(t)$ increases strictly, (8.24) is not greater than $h(T)$, that is, there do not exist finite T^* and N^* that satisfy both (8.21) and (8.23). This shows that when $c_T = c_N$, the unit might be replaced only at time T, irrespective of the number of working times.

8.1.2 Periodic Replacement

Suppose that the unit undergoes minimal repair at each failure [1, p. 96]. We consider four models of periodic replacement where the unit is replaced at a planned time T or (1) at a number N of uses, (2) at a number of N of failures, (3) at a number N of type 1 failures, and (4) at a number N of unit 1 failures, whichever occurs first. Optimum T^* and N^* that minimize the expected cost rates for each model are derived.

(1) Number of Uses

Suppose that the unit undergoes minimal repair at failures. From the assumptions in (1) of Section 8.1.1 that the unit is replaced at a planned time T or at a planned number of N of uses, whichever occurs first, the mean time to replacement is

$$T[1 - G^{(N)}(T)] + \int_0^T t \, dG^{(N)}(t) = \int_0^T [1 - G^{(N)}(t)] \, dt. \qquad (8.25)$$

Furthermore, let $H(t) \equiv \int_0^t h(u) \, du$ and $H_j \equiv \sum_{i=1}^j h_i$ $(j = 1, 2, \dots)$ be the cumulative hazard functions of $h(t)$ and h_j, i.e., $H(t)$ and H_j represent the expected numbers of failures caused by continuous deterioration with age during $[0, t]$ and by discrete number of uses until the jth number, respectively. Then, the expected number of failures when the unit is replaced at number N is

$$\int_0^T [H(t) + H_N] \, dG^{(N)}(t). \qquad (8.26)$$

Similarly, the expected number of failures when the unit is replaced at time T is

$$\sum_{j=0}^{N-1} [H(T) + H_j][G^{(j)}(T) - G^{(j+1)}(T)], \qquad (8.27)$$

where $H_0 \equiv 0$. Thus, by summing up (8.26) and (8.27), the expected number of failures before replacement is

$$\int_0^T h(t)[1 - G^{(N)}(t)] \, dt + \sum_{j=1}^N h_j G^{(j)}(T). \qquad (8.28)$$

Introduce the following costs: Cost c_M is the cost for minimal repair at each failure, and c_T and c_N are the respective replacement costs at time T and number N. Then, the expected cost rate is, from (8.25) and (8.28),

$$C_1(T, N) = \frac{c_M \left\{ \int_0^T h(t)[1 - G^{(N)}(t)] dt + \sum_{j=1}^N h_j G^{(j)}(T) \right\} + c_T + (c_N - c_T) G^{(N)}(T)}{\int_0^T [1 - G^{(N)}(t)] dt}. \qquad (8.29)$$

In particular, when the unit is replaced only at time T,

$$C_1(T) \equiv \lim_{N \to \infty} C_1(T, N) = \frac{c_M [H(T) + \sum_{j=1}^\infty h_j G^{(j)}(T)] + c_T}{T}, \qquad (8.30)$$

that agrees with (4.16) of [1, p. 102] for $h_j \equiv 0$. When the unit is replaced only at number N,

$$C_1(N) \equiv \lim_{T \to \infty} C_1(T, N) = \frac{c_M \left\{ \int_0^\infty h(t)[1 - G^{(N)}(t)]dt + H_N \right\} + c_N}{\int_0^\infty [1 - G^{(N)}(t)]dt},$$

$$(8.31)$$

that agrees with (4.33) of [1, p. 108] for $h(t) \equiv 0$.

We find optimum T^* and N^* that minimize $C_1(T, N)$ in (8.29) when $c_T = c_N$ and $G(t) = 1 - e^{-\theta t}$, i.e., $G^{(N)}(t) = \sum_{j=N}^\infty [(\theta t)^j / j!]e^{-\theta t}$ and $g^{(j)}(t) \equiv dG^{(j)}(t)/dt = \theta[(\theta t)^{j-1}/(j-1)!]e^{-\theta t}$ [155]. It is assumed that $h(t)$ increases strictly to ∞ and h_j increases strictly. Differentiating $C_1(T, N)$ in (8.29) with respect to T and setting it equal to zero,

$$Q_1(T; N) = \frac{c_T}{c_M}, \qquad (8.32)$$

where

$$Q_1(T; N) \equiv \left[h(T) + \frac{\sum_{j=1}^N h_j g^{(j)}(T)}{1 - G^{(N)}(T)} \right] \int_0^T [1 - G^{(N)}(t)] \, dt$$

$$- \int_0^T h(t)[1 - G^{(N)}(t)] \, dt - \sum_{j=1}^N h_j G^{(j)}(T).$$

First, prove that when h_j increases strictly,

$$\frac{\sum_{j=0}^N h_{j+1}[(\theta T)^j / j!]}{\sum_{j=0}^N [(\theta T)^j / j!]} \qquad (8.33)$$

also increases strictly with T and converges to h_{N+1} as $T \to \infty$ for any $N \geq 1$ [155]. Differentiating (8.33) with respect to T,

$$\frac{\theta}{\left\{ \sum_{j=0}^N [(\theta T)^j / j!] \right\}^2} \left[\sum_{j=1}^N h_{j+1} \frac{(\theta T)^{j-1}}{(j-1)!} \sum_{j=0}^N \frac{(\theta T)^j}{j!} - \sum_{j=0}^N h_{j+1} \frac{(\theta T)^j}{j!} \sum_{j=1}^N \frac{(\theta T)^{j-1}}{(j-1)!} \right].$$

The expression within the bracket of the numerator is

$$\sum_{j=1}^N \frac{(\theta T)^{j-1}}{(j-1)!} \sum_{i=0}^N \frac{(\theta T)^i}{i!} (h_{j+1} - h_{i+1})$$

$$= \sum_{j=1}^N \frac{(\theta T)^{j-1}}{(j-1)!} \sum_{i=0}^{j-1} \frac{(\theta T)^i}{i!} (h_{j+1} - h_{i+1}) - \sum_{j=1}^N \frac{(\theta T)^{j-1}}{(j-1)!} \sum_{i=j}^N \frac{(\theta T)^i}{i!} (h_{i+1} - h_{j+1})$$

$$= \sum_{j=1}^N \frac{(\theta T)^{j-1}}{j!} \sum_{i=0}^{j-1} \frac{(\theta T)^i}{i!} (h_{j+1} - h_{i+1})(j - i) > 0,$$

that implies that (8.33) increases strictly with T. Furthermore, it can be clearly seen that (8.33) tends to h_{N+1} as $T \to \infty$.

From the above results, $\lim_{T\to 0} Q_1(T; N) = 0$, $\lim_{T\to\infty} Q_1(T; N) = \infty$, and $Q_1(T; N)$ increases strictly with T. Thus, there exists a finite and unique T^* that satisfies (8.32) for any $N \geq 1$, and the resulting cost rate is

$$C_1(T^*, N) = c_M \left\{ h(T^*) + \frac{\theta \sum_{j=0}^{N-1} h_{j+1}[(\theta T^*)^j/j!]}{\sum_{j=0}^{N-1}[(\theta T^*)^j/j!]} \right\}. \tag{8.34}$$

Next, from the inequality $C_1(T, N+1) - C_1(T, N) > 0$ for a fixed $T > 0$,

$$L_1(N; T) \geq \frac{c_T}{c_M} \qquad (N = 1, 2, \dots), \tag{8.35}$$

where

$$L_1(N; T) \equiv \frac{\int_0^T [1 - G^{(N)}(t)]dt}{\int_0^T [G^{(N)}(t) - G^{(N+1)}(t)]dt} \left\{ \int_0^T h(t)[G^{(N)}(t) - G^{(N+1)}(t)]\, dt \right.$$
$$\left. + h_{N+1} G^{(N+1)}(T) \right\} - \int_0^T h(t)[1 - G^{(N)}(t)]\, dt - \sum_{j=1}^{N} h_j G^{(j)}(T).$$

Second, prove that when $h(t)$ increases strictly,

$$\frac{\int_0^T h(t)(\theta t)^N e^{-\theta t}dt}{\int_0^T (\theta t)^N e^{-\theta t}dt} \tag{8.36}$$

also increases strictly with N and converges to $h(T)$ as $N \to \infty$ for all $T > 0$ [155]. Denoting

$$q(T) \equiv \int_0^T h(t)(\theta t)^{N+1} e^{-\theta t}\, dt \int_0^T (\theta t)^N e^{-\theta t}\, dt$$
$$- \int_0^T h(t)(\theta t)^N e^{-\theta t}\, dt \int_0^T (\theta t)^{N+1} e^{-\theta t}\, dt,$$

it is clearly shown that $\lim_{T\to 0} q(T) = 0$, and

$$\frac{dq(T)}{dT} = (\theta T)^N e^{-\theta T} \int_0^T (\theta t)^N e^{-\theta t}(\theta T - \theta t)[h(T) - h(t)]\, dt > 0.$$

Thus, $q(T)$ increases strictly with T from 0 for any $N \geq 0$, that implies that (8.36) increases strictly with N. Furthermore, from the assumption that $h(t)$ increases,

$$\frac{\int_0^T h(t)(\theta t)^N e^{-\theta t}dt}{\int_0^T (\theta t)^N e^{-\theta t}dt} \leq h(T).$$

On the other hand, for any $\delta \in (0, T)$,

$$\frac{\int_0^T h(t)(\theta t)^N e^{-\theta t} dt}{\int_0^T (\theta t)^N e^{-\theta t} dt} = \frac{\int_0^{T-\delta} h(t)(\theta t)^N e^{-\theta t} dt + \int_{T-\delta}^T h(t)(\theta t)^N e^{-\theta t} dt}{\int_0^{T-\delta} (\theta t)^N e^{-\theta t} dt + \int_{T-\delta}^T (\theta t)^N e^{-\theta t} dt}$$

$$\geq \frac{h(T-\delta) \int_{T-\delta}^T h(t)(\theta t)^N e^{-\theta t} dt}{\int_0^{T-\delta} (\theta t)^N e^{-\theta t} dt + \int_{T-\delta}^T (\theta t)^N e^{-\theta t} dt}$$

$$= \frac{h(T-\delta)}{1 + \left[\int_0^{T-\delta} (\theta t)^N e^{-\theta t} dt / \int_{T-\delta}^T (\theta t)^N e^{-\theta t} dt \right]}.$$

The quantity in the bracket of the denominator is

$$\frac{\int_0^{T-\delta} (\theta t)^N e^{-\theta t} dt}{\int_{T-\delta}^T (\theta t)^N e^{-\theta t} dt} \leq \frac{e^{\theta T}}{\delta} \int_0^{T-\delta} \left(\frac{t}{T-\delta} \right)^N dt \to 0 \quad \text{as } N \to \infty.$$

Thus, it follows that

$$h(T-\delta) \leq \lim_{N \to \infty} \frac{\int_0^T h(t)(\theta t)^N e^{-\theta t} dt}{\int_0^T (\theta t)^N e^{-\theta t} dt} \leq h(T),$$

that completes the proof because δ is arbitrary.

From the above results,

$$L_1(N+1;T) - L_1(N;T) = \int_0^T [1 - G^{(N+1)}(t)] \, dt$$

$$\times \left[\theta(h_{N+2} - h_{N+1}) + \frac{\int_0^T h(t)(\theta t)^{N+1} e^{-\theta t} dt}{\int_0^T (\theta t)^{N+1} e^{-\theta t} dt} - \frac{\int_0^T h(t)(\theta t)^N e^{-\theta t} dt}{\int_0^T (\theta t)^N e^{-\theta t} dt} \right] > 0.$$

In addition, because T and N have to satisfy (8.32) and (8.35), the inequality (8.35) can be rewritten as

$$\theta \left\{ h_{N+1} - \frac{\sum_{j=0}^{N-1} h_{j+1}[(\theta T)^j / j!]}{\sum_{j=0}^{N-1} [(\theta T)^j / j!]} \right\} + \frac{\int_0^T h(t)(\theta t)^N e^{-\theta t} dt}{\int_0^T (\theta t)^N e^{-\theta t} dt} > h(T). \quad (8.37)$$

It is easily seen from (8.33) and (8.36) that there exists a finite and unique minimum $N^*(1 \leq N^* < \infty)$ that satisfies (8.37).

Example 8.2. We compute the optimum T^* and N^* that minimize the expected cost rate $C_1(T, N)$, using the computing procedure shown in Sect. 8.1.1. Suppose that $\theta = 1$, $h_j = (jp^2)/(q+jp)$, and $h(t) = 2\lambda t$, where h_j is the failure rate of a negative binomial distribution $p_j = jp^2 q^{j-1}$ $(j = 1, 2, \dots)$, $q \equiv 1-p$, and $h(t)$ is that of a Weibull distribution $[1 - \exp(-\lambda t^2)]$. Then, finite T^* and N^* exist uniquely and are computed numerically by solving the following equations:

Table 8.2. Optimum T^* and N^*, and expected cost rate $C_1(T^*, N^*)/c_M$ when $h_j = jp^2/(q + jp)$, $p = 0.05$, $h(t) = 2\lambda t$, and $\lambda = \pi p^2/[4(1 + q)^2]$

c_T/c_M	T^*	N^*	$C_1(T^*, N^*)/c_M$
0.1	10.9	10	0.0266
0.5	24.0	27	0.0518
1.0	35.4	43	0.0689
2.0	53.2	69	0.0918
3.0	67.2	91	0.1084
4.0	79.1	109	0.1220
5.0	89.7	126	0.1339
10.0	132.0	145	0.1779

$$\left[2\lambda T + \frac{p^2 \sum_{j=0}^{N-1} \{(j+1)T^j/[(1+jp)j!]\}}{\sum_{j=0}^{N-1}(T^j/j!)} \right] \sum_{j=0}^{N-1} \left(1 - \sum_{i=0}^{j} \frac{T^i}{i!} e^{-T} \right)$$

$$- (2\lambda) \sum_{j=0}^{N-1} (j+1) \left(1 - \sum_{i=0}^{j+1} \frac{T^i e^{-T}}{i!} \right) - p^2 \sum_{j=0}^{N-1} \frac{j+1}{1+jp} \left(1 - \sum_{i=0}^{j} \frac{T^i e^{-T}}{i!} \right) = \frac{c_T}{c_M},$$

$$\frac{(N+1)p^2}{Np+1} + \frac{(2\lambda)(N+1)\left[1 - \sum_{j=0}^{N+1}(T^j/j!)e^{-T} \right]}{1 - \sum_{j=0}^{N}(T^j/j!)e^{-T}}$$

$$- \frac{p^2 \sum_{j=0}^{N-1}\{(j+1)T^j/[(1+jp)j!]\}}{\sum_{j=0}^{N-1}(T^j/j!)} > 2\lambda T.$$

In addition, we set $\lambda = (\pi p^2)/[4(1 + q)^2]$, the same assumption as in Example 8.1. Table 8.2 presents the optimum T^*, N^*, and the expected cost rate $C_1(T^*, N^*)/c_M$ for c_T/c_M when $p = 0.05$, *i.e.*, the mean failure time is $(1 + q)/p = 39$. For example, when $c_T/c_M = 2.0$, the unit should be replaced at 53 units of time or at 69 uses, where one unit of time represents the mean time between uses. It is of interest that if uses occur constantly at a mean interval and c_T/c_M becomes large, then the unit may be replaced only at time T because $T^* < N^*$ for $c_T/c_M \geq 0.5$. ∎

(2) Nth Failure

The unit begins to operate at time 0 and undergoes only minimal repair at failures. Suppose that the unit is replaced at time T or at the Nth ($N = 1, 2, \ldots$) failure, whichever occurs first. Then, because the probability that j

failures occur exactly during $[0, t]$ is $p_j(t) \equiv \{[H(t)]^j/j!\}\, e^{-H(t)}$ [1, p. 97], the mean time to replacement is

$$T \sum_{j=0}^{N-1} p_j(T) + \int_0^T t\, h(t) p_{N-1}(t)\, dt = \sum_{j=0}^{N-1} \int_0^T p_j(t)\, dt,$$

and the expected number of failures before replacement is

$$\sum_{j=0}^{N-1} j p_j(T) + (N-1) \sum_{j=N}^{\infty} p_j(T) = N - 1 - \sum_{j=0}^{N-1}(N-1-j) p_j(T).$$

Thus, the expected cost rate is

$$C_2(T, N) = \frac{c_M[N-1-\sum_{j=0}^{N-1}(N-1-j)p_j(T)] + c_N + (c_T - c_N)\sum_{j=0}^{N-1} p_j(T)}{\sum_{j=0}^{N-1} \int_0^T p_j(t)dt}, \tag{8.38}$$

where the above costs are given in (8.29).

In particular, when the unit is replaced only at time T,

$$C_2(T) \equiv \lim_{N \to \infty} C_2(T, N) = \frac{c_M H(T) + c_T}{T}, \tag{8.39}$$

that agrees with (4.16) of [1] for the standard periodic replacement with a planned time T, and when it is replaced only at failure N,

$$C_2(N) \equiv \lim_{T \to \infty} C_2(T, N) = \frac{(N-1)c_M + c_N}{\sum_{j=0}^{N-1} \int_0^\infty p_j(t)dt}, \tag{8.40}$$

that agrees with (4.25) of [1].

It is assumed that the failure rate $h(t)$ increases strictly to ∞, and $c_T \leq c_N \leq c_M + c_T$ because the replacement cost at failure N would be higher than that at time T and lower than the total cost of minimal repair and the replacement at time T. In addition, let T_0 be the optimum time that minimizes $C_2(T)$ in (8.39). From the assumption that $h(t)$ increases strictly to ∞, there exists a finite and unique T_0 $(0 < T_0 < \infty)$ that satisfies [1, p. 102, 2]

$$Th(T) - H(T) = \frac{c_T}{c_M}. \tag{8.41}$$

Under the above conditions, we seek an optimum number N^* that minimizes $C_2(T, N)$ in (8.38) for a fixed T $(0 < T \leq \infty)$.

First, prove that [154]:

(1) If $h(t)$ increases, then $\int_0^\infty p_j(t)dt$ $(j = 0, 1, 2, \dots)$ decreases with j and converges to $1/h(\infty)$ as $j \to \infty$, and $\sum_{i=j+1}^{\infty} p_i(T)/\int_0^T p_j(t)dt$ increases with j and converges to $h(T)$ as $j \to \infty$ for any $T > 0$.

(2) $p_j(T)/\int_0^T p_j(t)dt$ increases with j and diverges to ∞ as $j \to \infty$ for any $T > 0$.

Using the relation

$$\frac{[H(t)]^j}{j!} = \int_0^t \frac{[H(u)]^{j-1}}{(j-1)!} h(u)\, du$$

and from the assumption that $h(t)$ increases, it follows that

$$\int_0^\infty p_j(t)\, dt = \int_0^\infty \left\{ \int_0^t \frac{[H(u)]^{j-1}}{(j-1)!} h(u)\, du \right\} e^{-H(t)}\, dt$$

$$\le \int_0^\infty \left\{ \int_0^t \frac{[H(u)]^{j-1}}{(j-1)!}\, du \right\} h(t) e^{-H(t)}\, dt$$

$$= \int_0^\infty p_{j-1}(t)\, dt,$$

and hence, $\int_0^\infty p_j(t)dt$ decreases with j. Furthermore,

$$\int_0^\infty p_j(t)\, dt = \int_0^\infty \frac{[H(t)]^j}{j!} \frac{h(t)}{h(t)} e^{-H(t)}\, dt$$

$$\ge \frac{1}{h(\infty)} \int_0^\infty \frac{[H(t)]^j}{j!} e^{-H(t)} h(t)\, dt = \frac{1}{h(\infty)}.$$

On the other hand, for any $T \in (0, \infty)$,

$$\int_0^\infty p_j(t)\, dt = \int_0^T p_j(t)\, dt + \int_T^\infty p_j(t)\, dt \le \int_0^T p_j(t)\, dt + \frac{1}{h(T)}.$$

Thus, because

$$\lim_{j \to \infty} \int_0^\infty p_j(t)\, dt \le \frac{1}{h(T)}$$

and T is arbitrary,

$$\lim_{j \to \infty} \int_0^\infty p_j(t)\, dt = \frac{1}{h(\infty)}.$$

Next, using the relation

$$\sum_{i=j+1}^\infty p_i(T) = \int_0^T \frac{[H(t)]^j}{j!} e^{-H(t)} h(t)\, dt,$$

we prove that

$$\frac{\int_0^T [H(t)]^j e^{-H(t)} h(t)dt}{\int_0^T [H(t)]^j e^{-H(t)} dt} \tag{8.42}$$

increases with j and converges to $h(T)$ as $j \to \infty$. Let us denote

$$q(T) \equiv \int_0^T [H(t)]^{j+1} e^{-H(t)} h(t)\, dt \int_0^T [H(t)]^j e^{-H(t)}\, dt$$

$$- \int_0^T [H(t)]^j e^{-H(t)} h(t)\, dt \int_0^T [H(t)]^{j+1} e^{-H(t)}\, dt.$$

Then, it is easily seen that $q(0) = 0$ and

$$\frac{dq(T)}{dT} = [H(T)]^j e^{-H(T)} \int_0^T [H(t)]^j e^{-H(t)} h(t)[H(T)-H(t)][h(T)-h(t)]\, dt \geq 0.$$

Thus, $q(T) > 0$ for all $T > 0$, and hence, (8.42) increases with j. Furthermore, it is clear that

$$\frac{\int_0^T [H(t)]^j e^{-H(t)} h(t)\, dt}{\int_0^T [H(t)]^j e^{-H(t)}\, dt} \leq h(T).$$

On the other hand, for any $T_0 \in (0, T)$,

$$\frac{\int_0^T [H(t)]^j e^{-H(t)} h(t)\, dt}{\int_0^T [H(t)]^j e^{-H(t)}\, dt} = \frac{\int_0^{T_0} [H(t)]^j e^{-H(t)} h(t)\, dt + \int_{T_0}^T [H(t)]^j e^{-H(t)} h(t)\, dt}{\int_0^{T_0} [H(t)]^j e^{-H(t)}\, dt + \int_{T_0}^T [H(t)]^j e^{-H(t)}\, dt}$$

$$\geq \frac{h(T_0) \int_{T_0}^T [H(t)]^j e^{-H(t)}\, dt}{\int_0^{T_0} [H(t)]^j e^{-H(t)}\, dt + \int_{T_0}^T [H(t)]^j e^{-H(t)}\, dt}$$

$$= \frac{h(T_0)}{1 + \left\{ \int_0^{T_0} [H(t)]^j e^{-H(t)}\, dt \big/ \int_{T_0}^T [H(t)]^j e^{-H(t)}\, dt \right\}}.$$

The bracket of the denominator is

$$\frac{\int_0^{T_0} [H(t)]^j e^{-H(t)}\, dt}{\int_{T_0}^T [H(t)]^j e^{-H(t)}\, dt} \leq \frac{\int_0^{T_0} e^{-H(t)}\, dt}{\int_{T_0}^T [H(t)/H(T_0)]^j e^{-H(t)}\, dt} \to 0 \quad \text{as } j \to \infty.$$

Thus,

$$h(T) \geq \frac{\int_0^T [H(t)]^j e^{-H(t)} h(t)\, dt}{\int_0^T [H(t)]^j e^{-H(t)}\, dt} \geq h(T_0)$$

implies that (8.42) tends to $h(T)$ as $j \to \infty$ because T_0 is arbitrary.

Similarly, we prove **(2)** as follows: Because $H(T) > H(t)$ for $T > t$,

$$\frac{p_{j+1}(T)}{\int_0^T p_{j+1}(t)\, dt} - \frac{p_j(T)}{\int_0^T p_j(t)\, dt} = \frac{p_{j+1}(T)}{\int_0^T p_{j+1}(t)\, dt \int_0^T p_j(t)\, dt} \int_0^T p_j(t) \left[1 - \frac{H(t)}{H(T)} \right] dt > 0,$$

and

$$\frac{p_j(T)}{\int_0^T p_j(t)\, dt} = \frac{e^{-H(T)}}{\int_0^T [H(t)/H(T)]^j e^{-H(t)}\, dt} \to \infty \quad \text{as } j \to \infty.$$

Using the above results in **(1)** and **(2)**, we have the following optimum number N^* that minimizes $C_2(T, N)$ in (8.38) for a fixed T $(0 < T < \infty)$:

(i) If $c_N = c_M + c_T$ and $T > T_0$, then there exists a finite and unique minimum N^* that satisfies

$$L_2(N;T) \geq c_T \qquad (N = 1, 2, \dots), \qquad (8.43)$$

where

$$L_2(N;T) \equiv \frac{\sum_{j=0}^{N-1} \int_0^T p_j(t)dt}{\int_0^T p_N(t)dt} \left[c_M \sum_{j=N}^{\infty} p_j(T) - (c_N - c_T)p_N(T) \right]$$

$$- c_M \left[N - 1 \sum_{j=0}^{N-1} (N - 1 - j)p_j(T) \right] - (c_N - c_T) \sum_{j=N}^{\infty} p_j(T),$$

and $\sum_{j=0}^{-1} \equiv 0$, and the resulting cost rate is

$$\frac{\sum_{j=N^*}^{\infty} p_j(T)}{\int_0^T p_{N^*-1}(t)dt} < \frac{C_2(T, N^*)}{c_M} \leq \frac{\sum_{j=N^*+1}^{\infty} p_j(T)}{\int_0^T p_{N^*}(t)dt}. \qquad (8.44)$$

(ii) If $c_N = c_M + c_T$ and $T \leq T_0$, then $N^* = \infty$, *i.e.*, the unit is replaced only at time T, and the expected cost rate is given in (8.39).

(iii) If $c_N < c_M + c_T$, then there exists a finite and unique N^* that satisfies (8.43), and

$$\frac{c_M \sum_{j=N^*-1}^{\infty} p_j(T) - (c_N - c_T)p_{N^*-1}(T)}{\int_0^T p_{N^*-1}(t)dt}$$

$$< C_2(T, N^*) \leq \frac{c_M \sum_{j=N^*}^{\infty} p_j(T) - (c_N - c_T)p_{N^*}(T^*)}{\int_0^T p_{N^*}(t)dt}. \qquad (8.45)$$

We prove (i), (ii), and (iii) as follows: The inequalities $C_2(T, N+1) - C_2(T, N)$ implies (8.43). First, assume that $c_N = c_M + c_T$. Then, using the relation

$$\int_0^T p_N(t)h(t)\, dt = \sum_{j=N+1}^{\infty} p_j(T)$$

or

$$\int_T^{\infty} p_N(t)h(t)\, dt = \sum_{j=0}^{N} p_j(T),$$

we can easily see from **(1)** that

$$L_2(N+1, T) - L_2(N, T)$$

$$= c_M \sum_{j=0}^{N} \int_0^T p_j(t)\, dt \left[\frac{\sum_{j=N+2}^{\infty} p_j(T)}{\int_0^T p_{N+1}(t)dt} - \frac{\sum_{j=N+1}^{\infty} p_j(T)}{\int_0^T p_N(t)dt} \right] > 0$$

and

$$L_2(\infty, T) \equiv \lim_{N \to \infty} L(N, T) = c_M[Th(T) - H(T)],$$

that is equal to the left-hand side of (8.41). Thus, from the notation T_0 that satisfies (8.41), if $L_2(\infty, T) > c_T/c_M$, i.e., $T > T_0$, then there exists a finite and unique N^* that satisfies (8.43). In addition, substituting the inequality (8.43) in (8.38), we easily get (8.44). On the other hand, if $L_2(\infty, T) \le c_T/c_M$, i.e., $T \le T_0$, then $C_2(T, N)$ decreases with N, and hence, $N^* = \infty$.

Next, assume that $c_N < c_M + c_T$. Then, from **(1)** and **(2)**,

$$L_2(N+1, T) - L_2(N, T) =$$

$$\sum_{j=0}^{N} \int_0^T p_j(t)\, dt \left\{ c_M \left[\frac{\sum_{j=N+2}^{\infty} p_j(T)}{\int_0^T p_{N+1}(t)dt} - \frac{\sum_{j=N+1}^{\infty} p_j(T)}{\int_0^T p_N(t)dt} \right] \right.$$

$$\left. + (c_M - c_N + c_T) \left[\frac{p_{N+1}(T)}{\int_0^T p_{N+1}(t)dt} - \frac{p_N(T)}{\int_0^T p_N(t)dt} \right] \right\} > 0, \qquad (8.46)$$

and $\lim_{N \to \infty} L_2(N, T) = \infty$, that complete the results (i), (ii), and (iii).

Example 8.3.   It is very difficult to discuss analytically both optimum number N^* and time T^* that minimize $C_2(T, N)$ in (8.38). Consider the particular case where the failure time of the unit has a Weibull distribution, *i.e.*, $H(t) = t^m$ and $h(t) = mt^{m-1}$ for $m > 1$. Then, the optimum policies are:

(iv) If $c_N \ge c_M + c_T$, then the unit is replaced only at time

$$T^* = \left[\frac{c_T}{c_M(m-1)} \right]^{1/m}.$$

(v) If $c_T < c_N < c_M + c_T$, then the unit is replaced at failure N^* or at time T^*, whichever occurs first, where N^* is the unique minimum that satisfies (8.43) and T^* satisfies

$$Q_2(T^*, N^*) = c_T, \qquad (8.47)$$

where

$$Q_2(T, N) \equiv h(T) \sum_{j=0}^{N-1} \int_0^T p_j(t)\, dt \left[c_M - (c_M - c_N + c_T) \frac{p_{N-1}(T)}{\sum_{j=0}^{N-1} p_j(T)} \right]$$

$$- c_M \left[N - 1 - \sum_{j=0}^{N-1} (N - 1 - j)\, p_j(T) \right] - (c_N - c_T) \sum_{j=N}^{\infty} p_j(T),$$

and

$$\lim_{N \to \infty} Q_2(T, N) = c_M[Th(T) - H(T)].$$

(vi) If $c_N \leq c_T$, then the unit is replaced only at failure

$$N^* = \left[\frac{c_N - c_M}{c_M(m-1)}\right] + 1,$$

where $[x]$ denotes the greatest integer contained in x.

We prove the result (iv), (v), and (vi): Differentiating $C_2(T, N)$ in (8.38) with respect to T and setting it equal to zero, we have (8.47). It can be clearly seen that $Q_2(0, N) = 0$ and

$$Q_2(\infty, N) \equiv \lim_{T \to \infty} Q_2(T, N)$$

$$= (c_N - c_T)h(\infty) \sum_{j=0}^{N-1} \int_0^\infty p_j(t)\,dt - [c_M(N-1) + c_N - c_T]. \quad (8.48)$$

A necessary condition that finite N^* and T^* minimize $C_2(T, N)$ is that they satisfy (8.43) and (8.47), respectively.

(vii) When $c_N \geq c_M + c_T$, there exists a finite T^* that satisfies (8.47) from (8.48). In addition, from **(1)** and **(2)**,

$$L_2(N, T^*) - c_T = L_2(N, T^*) - Q_2(T^*, N)$$

$$= \frac{\sum_{j=0}^{N-1} \int_0^{T^*} p_j(t)dt}{\int_0^{T^*} p_N(t)dt}\left[c_M \sum_{j=N}^{\infty} p_j(T^*) - (c_N - c_T)p_N(T^*)\right]$$

$$- h(T^*)\sum_{j=0}^{N-1}\int_0^{T^*} p_j(t)\,dt$$

$$\times \left[c_M - (c_M - c_N + c_T)\frac{p_{N-1}(T^*)}{\sum_{j=0}^{N-1} p_j(T^*)}\right]$$

$$\leq c_M \sum_{j=0}^{N-1}\int_0^{T^*} p_j(t)\,dt\left[\frac{\sum_{j=N+1}^{\infty} p_j(T^*)}{\int_0^{T^*} p_N(t)dt} - h(T^*)\right] < 0,$$

that implies that $N^* = \infty$, i.e., we should replace the unit only at time T_0.

(viii) When $c_T < c_N < c_M + c_T$, from (8.46), $L_2(N, T)$ increases with N and $\lim_{N \to \infty} L_2(N, T) = \infty$, and hence, there exists a finite and unique minimum N^* that satisfies (8.43) for all $T > 0$. In addition, from (8.48), a finite T^* that satisfies (8.47) exists for all N because $h(\infty) = \infty$ for $m > 1$. Thus, there exist finite N^* and T^* that satisfy (8.43) and (8.47), respectively.

(ix) When $c_N \leq c_T$, $L_2(N, T)$ also increases with N to ∞. Thus, there exists a finite and unique N^* that satisfies (8.43), and $L_2(N^* - 1, T) < c_T$. Thus,

$$Q_2(T, N^*) - c_T < Q_2(T, N^*) - L_2(N^* - 1, T)$$

$$= \sum_{j=0}^{N^*-1} \int_0^T p_j(t)\, dt \left\{ h(T)[c_M - (c_M - c_N + c_T)] \frac{p_{N^*-1}(T)}{\sum_{j=0}^{N^*=1} p_j(T)} \right.$$

$$\left. - \frac{c_M \sum_{j=N^*}^{\infty} p_j(T) + (c_M - c_N + c_T)p_{N^*-1}(T)}{\int_0^T p_{N^*-1}(t)dt} \right\}$$

$$\leq c_M \sum_{j=0}^{N^*-1} \int_0^T p_j(t)\, dt \left[h(T) \frac{\sum_{j=0}^{N^*-2} p_j(T)}{\sum_{j=0}^{N^*-1} p_j(T)} - \frac{\sum_{j=N^*-1}^{\infty} p_j(T)}{\int_0^T p_{N^*-1}(t)dt} \right]$$

$$\leq c_M \frac{\sum_{j=0}^{N^*-1} \int_0^T p_j(t)dt \sum_{j=N^*-1}^{\infty} p_j(T)}{\int_T^{\infty} p_{N^*-1}(t)dt \int_0^T p_{N^*-1}(t)dt}$$

$$\times \left[\frac{\int_0^T p_{N^*-1}(t)dt}{\sum_{j=N^*-1}^{\infty} p_j(T)} - \int_0^{\infty} p_{N^*-1}(t)\, dt \right].$$

Let us denote

$$K(T, N) \equiv \frac{\int_0^T p_{N-1}(t)dt}{\sum_{j=N-1}^{\infty} p_j(T)} = \frac{\int_0^T p_{N-1}(t)dt}{\int_0^T p_{N-2}(t)h(t)dt}.$$

for a fixed N ($1 \leq N < \infty$). Then, from the assumption of $H(t) = t^m$,

$$K(0, N) \equiv \lim_{T \to 0} K(T, N) = \lim_{T \to 0} \frac{T}{m(N-1)} = 0,$$

$$K(\infty, N) \equiv \lim_{T \to \infty} K(T, N) = \int_0^{\infty} p_{N-1}(t)\, dt,$$

$$\frac{dK(T, N)}{dT} > 0.$$

Thus, $Q_2(T, N^*) < c_T$ for all $T > 0$, i.e., the unit is replaced only at the N^*th failure that satisfies [153]

$$\frac{\sum_{j=0}^{N-1} \int_0^{\infty} p_j(t)dt}{\int_0^{\infty} p_N(t)dt} - (N-1) \geq \frac{c_N}{c_M} \qquad (N = 1, 2, \dots), \qquad (8.49)$$

and

$$\frac{1}{\int_0^{\infty} p_{N^*-1}(t)dt} < \frac{C_2(N^*)}{c_M} \leq \frac{1}{\int_0^{\infty} p_{N^*}(t)dt}. \qquad (8.50)$$

In the case of $c_T < c_N < c_M + c_T$, we can specify the computing procedure for obtaining optimum T^* and N^*:

1. Set $N_0 = \left[(c_N - c_M)/[c_M(m-1)]\right] + 1$ and compute T_1 to satisfy (8.47).
2. Set $T = T_1$ and compute N_1 to satisfy satisfy (8.43).
3. Set $N = N_1$ and compute T_2 to satisfy (8.47).

Table 8.3. Optimum number N^* and T^* when $H(t) = t^2$ and $c_M = 5$

c_N	$c_T = 10$		$c_T = 15$	
	N^*	T^*	N^*	T^*
10	2	∞	2	∞
11	3	6.99	2	∞
12	3	3.38	2	∞
13	4	2.27	2	∞
14	5	1.92	2	∞
15	∞	1.41	3	∞
16	∞	1.41	4	8.61
17	∞	1.41	4	4.11
18	∞	1.41	5	2.69
19	∞	1.41	6	2.28
20	∞	1.41	∞	1.73

4. Continue until $N_j = N_{j+1}(j = 0, 1, 2, \dots)$.

Table 8.3 presents the optimum N^* and T^* for $c_N = 10$–20 when $m = 2$, $c_T = 10, 15$, and $c_M = 5$. The optimum finite time T^* and number N^* exist for $11 \leq c_N \leq 14$ when $c_T = 10$, and $16 \leq c_N \leq 19$ when $c_T = 15$. It is of great interest that the optimum policy varies with maintenance costs. ∎

(3) Two Types of Failures

Consider a unit with two types of failures [1, p. 110]: When the unit fails, type 1 failure occurs with probability α $(0 < \alpha < 1)$ and is removed by minimal repair, and type 2 failure occurs with probability $1 - \alpha$ and is removed by replacement. Type 1 failure means a minor failure that is easily restored to the same operating state before failure by minimal repair, whereas type 2 failure is a total breakdown.

Suppose that the unit is replaced before failure at a planned time T or the Nth type 1 failure, whichever occurs first. Then, the probability that the unit is replaced at time T is

$$\sum_{j=0}^{N-1} \alpha^j p_j(T), \tag{8.51}$$

the probability that it is replaced at failure N of type 1 is

$$\alpha^N \int_0^T p_{N-1}(t)h(t)\, dt = \alpha^N \sum_{j=N}^{\infty} p_j(T), \tag{8.52}$$

and the probability that it is replaced at type 2 failure is

$$\sum_{j=1}^{N}(1-\alpha)\alpha^{j-1}\int_0^T p_{j-1}(t)h(t)\,dt = \sum_{j=1}^{N}(1-\alpha)\alpha^{j-1}\sum_{i=j}^{\infty}p_i(T), \qquad (8.53)$$

where it is noted that $(8.51)+(8.52)+(8.53)=1$.

Furthermore, the expected number of minimal repairs, $i.e.$, type 1 failures, before replacement is

$$\sum_{j=1}^{N-1}j\alpha^j p_j(T) + (N-1)\alpha^N \sum_{j=N}^{\infty}p_j(T) + \sum_{j=1}^{N}(j-1)(1-\alpha)\alpha^{j-1}\sum_{i=j}^{\infty}p_i(T)$$

$$=\sum_{j=1}^{N-1}\alpha^j \sum_{i=j}^{\infty}p_i(T). \qquad (8.54)$$

The mean time to replacement is

$$T\sum_{j=0}^{N-1}\alpha^j p_j(T) + \alpha^N \int_0^T t\,p_{N-1}(t)h(t)\,dt + \sum_{j=1}^{N}(1-\alpha)\alpha^{j-1}\int_0^T t\,p_{j-1}(t)h(t)\,dt$$

$$=\sum_{j=0}^{N-1}\alpha^j \int_0^T p_j(t)\,dt. \qquad (8.55)$$

Therefore, the expected cost rate is

$$C_1(T,N) = \frac{c_M\sum_{j=1}^{N-1}\alpha^j \sum_{i=j}^{\infty}p_i(T)+c_F}{\sum_{j=0}^{N-1}\alpha^j \int_0^T p_j(t)\,dt}, \qquad (8.56)$$

where $c_M = $ cost for minimal repair at type 1 failure, $c_T = $ cost for replacement at time T, $c_N = $ cost for replacement at failure N of type 1, and $c_F = $ cost for replacement at type 2 failure. In particular, when $\alpha = 1$, $C_1(T,N)$ agrees with (8.38). When $\alpha = 0$,

$$C_1(T,N) = \frac{c_F - (c_F - c_T)p_0(T)}{\int_0^T p_0(t)\,dt}, \qquad (8.57)$$

that corresponds to the expected cost rate for the standard age replacement model [1, p. 72] by replacing $p_0(t) = 1 - F(t)$.

When the unit is replaced before failure only at time T,

$$C_1(T) \equiv \lim_{N\to\infty} C_1(T,N)$$

$$=\frac{c_M\sum_{j=1}^{\infty}\alpha^j \sum_{i=j}^{\infty}p_i(T)+c_F-(c_F-c_T)\sum_{j=0}^{\infty}\alpha^j p_j(T)}{\sum_{j=0}^{\infty}\alpha^j \int_0^T p_j(t)\,dt}, \qquad (8.58)$$

and when the unit is replaced before failure only at failure N of type 1,

$$C_1(N) \equiv \lim_{T \to \infty} C_1(T, N) = \frac{c_M \sum_{j=1}^{N-1} \alpha^j + c_F - (c_F - c_T)\alpha^N}{\sum_{j=0}^{N-1} \alpha^j \int_0^\infty p_j(t)dt}. \tag{8.59}$$

The optimum policy that minimizes $C_1(N)$ was discussed analytically [155].

(4) Two Types of Units

Consider a system with two types of units that operate independently [1, p. 112]. When unit 1 fails, it undergoes minimal repair instantaneously and begins to operate again. When unit 2 fails, the system is replaced without repairing it, i.e., unit 2 is vital or essential, and unit 1 is nonvital or non-essential. Failures of unit 1 occur in a non-homogeneous Poisson process with $p_j(t)$ $(j = 0, 1, 2, \dots)$, and unit 2 has a failure distribution $G(t)$ with a finite mean $1/\theta < \infty$, where $\overline{G}(t) \equiv 1 - G(t)$.

Suppose that the system is replaced before failure at a planned time T or the Nth failure of unit 1, whichever occurs first. Then, the probability that the unit is replaced at time T is

$$\overline{G}(T) \sum_{j=0}^{N-1} p_j(T), \tag{8.60}$$

the probability that it is replaced at failure N of unit 1 is

$$\int_0^T \overline{G}(t)p_{N-1}(t)h(t)\,dt = 1 - \overline{G}(T) \sum_{j=0}^{N-1} p_j(T) - \sum_{j=0}^{N-1} \int_0^T p_j(t)\,dG(t), \tag{8.61}$$

and the probability that it is replaced at failure of unit 2 is

$$\sum_{j=0}^{N-1} \int_0^T p_j(t)\,dG(t). \tag{8.62}$$

Furthermore, the expected number of minimal repairs before replacement is

$$\overline{G}(T) \sum_{j=1}^{N-1} j p_j(T) + (N-1) \int_0^T \overline{G}(t)p_{N-1}(t)h(t)\,dt + \sum_{j=1}^{N-1} j \int_0^T p_j(t)\,dG(t)$$

$$= \sum_{j=1}^{N-1} \int_0^T \overline{G}(t)p_{j-1}(t)h(t)\,dt. \tag{8.63}$$

The mean time to replacement is

$$T\overline{G}(T)\sum_{j=0}^{N-1} p_j(T) + \int_0^T t\,\overline{G}(t)p_{N-1}(t)h(t)\,dt + \sum_{j=0}^{N-1}\int_0^T t\,p_j(t)\,dG(t)$$

$$= \sum_{j=0}^{N-1}\int_0^T \overline{G}(t)p_j(t)\,dt. \tag{8.64}$$

Therefore, the expected cos rate is

$$C_2(T,N) = \frac{c_M \sum_{j=0}^{N-2}\int_0^T \overline{G}(t)p_j(t)h(t)dt + c_N}{\sum_{j=0}^{N-1}\int_0^T \overline{G}(t)p_j(t)dt},$$

$$\frac{+(c_T - c_N)\overline{G}(T)\sum_{j=0}^{N-1} p_j(T) + (c_F - c_N)\sum_{j=0}^{N-1}\int_0^T p_j(t)dG(t)}{}$$

$$\tag{8.65}$$

where c_M = cost for minimal repair at unit 1 failure, c_T = cost for replacement at time T, c_N = cost for replacement at failure N of unit 1, and c_F = cost for replacement at unit 2 failure. In particular, when $\overline{G}(t) = 1$ for $t \geq 0$, i.e., unit 2 does not fail at all, $C_2(T,N)$ agrees with (8.38).

When the unit is replaced before failure only at time T,

$$C_2(T) \equiv \lim_{N\to\infty} C_2(T,N)$$

$$= \frac{c_M \int_0^T \overline{G}(t)h(t)dt + c_T + (c_F - c_T)G(T)}{\int_0^T \overline{G}(t)dt}, \tag{8.66}$$

and when the unit is replaced before failure only at failure N of unit 1,

$$C_2(N) \equiv \lim_{T\to\infty} C_2(T,N)$$

$$= \frac{c_M \sum_{j=0}^{N-2}\int_0^\infty \overline{G}(t)p_j(t)h(t)dt + c_N + (c_F - c_N)\sum_{j=0}^{N-1}\int_0^\infty p_j(t)dG(t)}{\sum_{j=0}^{N-1}\int_0^\infty \overline{G}(t)p_j(t)dt}. \tag{8.67}$$

The optimum policy that minimizes $C_2(N)$ was extensively discussed [153].

8.1.3 Block Replacement

When the unit has a failure distribution $F(t)$, it is replaced with a new one at each failure [1, p. 117]. Suppose that the unit is also replaced at time T or at the Nth ($N = 1, 2, \dots$) failure, whichever occurs first. Then, substituting $p_j(t)$ in (8.38) for $F^{(j)}(t) - F^{(j+1)}(t)$ formally,

$$C_3(T,N) = \frac{c_M \sum_{j=1}^{N-1} F^{(j)}(T) + c_T - (c_T - c_N)F^{(N)}(T)}{\int_0^T [1 - F^{(N)}(t)]dt}, \tag{8.68}$$

where c_M is the replacement cost at each failure, and c_T and c_N are given in (8.29). In general, it would be very difficult to discuss analytically optimum T^* and N^* that minimize $C_3(T,N)$.

In particular, when the unit is replaced only at time T,

$$C_3(T) \equiv \lim_{N \to \infty} C_3(T, N) = \frac{c_M M(T) + c_T}{T}, \qquad (8.69)$$

that agrees with (5.1) of [1] for the standard block replacement with a planned time T, and when it is replaced only at failure N,

$$C_3(N) \equiv \lim_{T \to \infty} C_3(T, N) = \frac{(N-1)c_M + c_N}{\int_0^\infty [1 - F^{(N)}(t)]dt}, \qquad (8.70)$$

where $M(t) \equiv \sum_{j=1}^\infty F^{(j)}(t)$ that represents the expected number of failures during $(0, t]$.

A variety of combined replacement models of age, periodic, and block replacements were proposed [1, p. 125].

8.2 Modified Replacement Policies

It might be wise to replace practically a unit at the completion of its working time even if T comes because it continues to work for some job. The unit might be working as long as possible in the case where the replacement cost after failure is not so high. In this case, the unit should be replaced at T or N, whichever occurs *last* rather than at T or N, whichever occurs *first*. From such viewpoints, we consider two modified replacement models proposed in Sect. 8.1. We would estimate replacement costs for each model and should discuss which model is better. Furthermore, we take up the replacement of a backup model that has been discussed in Chap. 7.

(1) Replacement over Time T

Suppose in **(2)** of 8.1.1 that the unit is replaced before failure at the Nth completion of working times or at the first completion over time T, whichever occurs first. Then, the probability that the unit is replaced at number N is

$$\int_0^T \overline{F}(t) \, dG^{(N)}(t), \qquad (8.71)$$

and the probability that it is replaced at the first completion of working times over time T is

$$\sum_{j=0}^{N-1} \int_0^T \left[\int_{T-t}^\infty \overline{F}(t + u) dG(u) \right] dG^{(j)}(t). \qquad (8.72)$$

Furthermore, the probability that the unit is replaced at failure before time T is

$$\sum_{j=0}^{N-1} \int_0^T [G^{(j)}(t) - G^{(j+1)}(t)] \, dF(t), \tag{8.73}$$

and it is replaced at failure after time T is

$$\sum_{j=0}^{N-1} \int_0^T \left\{ \int_{T-t}^\infty [F(t+u) - F(T)] \, dG(u) \right\} dG^{(j)}(t), \tag{8.74}$$

where note that $(8.71) + (8.72) + (8.73) + (8.74) = 1$. Thus, the mean time to replacement is

$$\int_0^T t \, \overline{F}(t) \, dG^{(N)}(t) + \int_0^T t \, [1 - G^{(N)}(t)] \, dF(t)$$

$$+ \sum_{j=0}^{N-1} \int_0^T \left\{ \int_{T-t}^\infty \left[\int_T^{t+u} v \, dF(v) \right] dG(u) \right\} dG^{(j)}(t)$$

$$+ \sum_{j=0}^{N-1} \int_0^T \left[\int_{T-t}^\infty (t+u) \overline{F}(t+u) \, dG(u) \right] dG^{(j)}(t)$$

$$= \int_0^T [1 - G^{(N)}(t)] \overline{F}(t) \, dt + \sum_{j=0}^{N-1} \int_0^T \int_T^\infty [\overline{G}(u-t) \overline{F}(u) \, du] \, dG^{(j)}(t). \tag{8.75}$$

Therefore, the expected cost rate is

$$C_2(T, N) = \frac{c_F - (c_F - c_T) \sum_{j=0}^{N-1} \int_0^T [\int_{T-t}^\infty F(t+u) dG(u)] dG^{(j)}(t)}{- (c_F - c_N) \int_0^T \overline{F}(t) dG^{(N)}(t)}{\int_0^T [1 - G^{(N)}(t)] \overline{F}(t) dt + \sum_{j=0}^{N-1} \int_0^T [\int_T^\infty \overline{G}(u-t) \overline{F}(u) du] dG^{(j)}(t)}, \tag{8.76}$$

where c_F = replacement cost at failure, c_T = replacement cost over time T, and c_N = replacement cost at number N. In particular, when the unit is replaced only at number N,

$$C_2(N) \equiv \lim_{T \to \infty} C_2(T, N) = \frac{c_F - (c_F - c_N) \int_0^\infty \overline{F}(t) dG^{(N)}(t)}{\int_0^\infty [1 - G^{(N)}(t)] \overline{F}(t) dt}, \tag{8.77}$$

that agrees with (8.19).

Next, suppose in (2) of 8.1.2 that the unit is replaced at the Nth failure or at the first failure over time T, whichever occurs first. Then, the probability that the unit is replaced at failure N is

$$\int_0^T p_{N-1}(t) h(t) \, dt = \sum_{j=N}^\infty p_j(T),$$

and the probability that it is replaced at the first failure over time T is

$$\left\{ 1 + \sum_{j=1}^{N-1} \int_0^T \frac{[H(t)]^{j-1}}{(j-1)!} h(t) \, dt \right\} \overline{F}(T) = \sum_{j=0}^{N-1} p_j(T).$$

Furthermore, the mean time to replacement is

$$\int_0^T t p_{N-1}(t) h(t) \, dt + \left\{ 1 + \sum_{j=1}^{N-1} \int_0^T \frac{[H(t)]^{j-1}}{(j-1)!} h(t) \, dt \right\} \int_T^\infty u \, dF(u)$$

$$= \sum_{j=0}^{N-1} \int_0^T p_j(t) \, dt + \sum_{j=0}^{N-1} \frac{[H(T)]^j}{j!} \int_T^\infty e^{-H(u)} \, du. \tag{8.78}$$

Because the expected number of failures before replacement is given in (8.38), the expected cost rate is

$$C_2(T, N) = \frac{c_M \left[N - 1 - \sum_{j=0}^{N-1} (N-1-j) p_j(T) \right] + c_N + (c_T - c_N) \sum_{j=0}^{N-1} p_j(T)}{\sum_{j=0}^{N-1} \int_0^T p_j(t) \, dt + \sum_{j=0}^{N-1} p_j(T) \int_T^\infty e^{-[H(t)-H(T)]} \, dt}. \tag{8.79}$$

When the unit is replaced only at failure N, the expected cost rate $C_2(\infty, N)$ agrees with (8.40).

(2) Replacement Whichever Occurs Last

Suppose in (2) of 8.1.1 that the unit is replaced before failure at a planned time T or at the Nth completion of working times, whichever occurs last. Then, the probability that the unit is replaced at number N is

$$\int_T^\infty \overline{F}(t) \, dG^{(N)}(t), \tag{8.80}$$

the probability that it is replaced at time T is

$$\overline{F}(T) G^{(N)}(T), \tag{8.81}$$

and the probability that it is replaced at failure is

$$F(T) + \int_T^\infty [1 - G^{(N)}(t)] \, dF(t), \tag{8.82}$$

where $(8.80) + (8.81) + (8.82) = 1$. Thus, the mean time to replacement is

$$\int_T^\infty t \, \overline{F}(t) \, dG^{(N)}(t) + T \overline{F}(T) G^{(N)}(T) + \int_0^T t \, dF(t) + \int_T^\infty t \, [1 - G^{(N)}(t)] \, dF(t)$$

$$= \int_0^T \overline{F}(t) \, dt + \int_T^\infty \overline{F}(t)[1 - G^{(N)}(t)] \, dt. \tag{8.83}$$

Therefore, the expected cost rate is

$$C_2(T, N) = \frac{c_F - (c_F - c_T)\overline{F}(T)G^{(N)}(T) - (c_F - c_N)\int_T^\infty \overline{F}(t)dG^{(N)}(t)}{\int_0^T \overline{F}(t)dt + \int_T^\infty \overline{F}(t)[1 - G^{(N)}(t)]dt}.$$

(8.84)

Compared with the expected cost rate in (8.17) when $c_T = c_N$, both numerator and denominator are larger than those in (8.17). In particular, when we set formally $N = 0$, i.e., $G^{(0)}(T) \equiv 1$ for $T \geq 0$, and 0 for $T < 0$, $C_2(T, 0)$ agrees with (8.18), and $\lim_{T \to 0} C_2(T, N)$ agrees with (8.19). Clearly,

$$\lim_{T \to \infty} C_2(T, N) = \lim_{N \to \infty} C_2(T, N) = \frac{c_F}{\mu}$$

that is the expected cost rate when the unit is replaced only at failure, where $\mu \equiv \int_0^\infty \overline{F}(t)dt$.

Next, suppose in **(2)** of 8.1.2 that the unit is replaced at time T or at the Nth failure, whichever occurs last. Then, because the probability that the unit is replaced at time T is $\sum_{j=N}^\infty p_j(T)$, and the probability that it is replaced at failure N is $\sum_{j=0}^{N-1} p_j(T)$, the mean time to replacement is

$$T \sum_{j=N}^\infty p_j(T) + \int_T^\infty t\, h(t)p_{N-1}(t)\, dt = T + \sum_{j=0}^{N-1} \int_T^\infty p_j(t)\, dt,$$

(8.85)

and the expected number of failures before replacement is

$$\sum_{j=N}^\infty j p_j(T) + (N - 1) \sum_{j=0}^{N-1} p_j(T) = N - 1 + \sum_{j=N}^\infty (j - N + 1)p_j(T).$$

(8.86)

Therefore, the expected cost rate is

$$C_2(T, N) = \frac{c_M[N - 1 + \sum_{j=N}^\infty (j - N + 1)p_j(T)] + c_T + (c_N - c_T)\sum_{j=0}^{N-1} p_j(T)}{T + \sum_{j=0}^{N-1} \int_T^\infty p_j(t)dt}.$$

(8.87)

In particular, when $T = 0$ and $N = 0$, i.e, $\sum_0^{-1} \equiv 0$, $C_2(T, N)$ agrees with (8.40) and (8.39), respectively. It would be important to compare analytically and numerically two expected cost rates $C_2(T, N)$ in (8.17) and (8.84), and $C_2(T, N)$ in (8.38) and (8.87).

(3) Backup Policy

We consider the following backup model: Tasks arrive at a counter and receive some service from a system according to a general distribution $G(t)$. When N tasks are completed, they are backed up and stored in storage. However, if the system fails before the completed services of N tasks, all completed tasks until failure are useless. This is applied to a computer system by replacing *service* with *processing* and to some production system by replacing *service* with *production*.

Arrival tasks are backed up at time T or the Nth completion of services, whichever occurs first. Introduce the following costs: c_T and c_N are the respective backup costs at time T and at number N. In addition, $c_F + c_D t$ is the cost when the system fails at time t because all tasks completed until time t are lost. This expected cost is, from **(2)** of Sect. 8.1.1,

$$\int_0^T (c_F + c_D t)[1 - G^{(N)}(t)] \, \mathrm{d}F(t). \tag{8.88}$$

Thus, by the similar method for obtaining (8.17), the expected cost rate is

$$C_2(T, N) = \frac{\begin{array}{c} c_F - (c_F - c_T)\overline{F}(T)[1 - G^{(N)}(T)] \\ - (c_F - c_N) \int_0^T \overline{F}(t)\mathrm{d}G^{(N)}(t) + c_D \int_0^T t[1 - G^{(N)}(t)]\mathrm{d}F(t) \end{array}}{\int_0^T [1 - G^{(N)}(t)]\overline{F}(t)\mathrm{d}t}. \tag{8.89}$$

When $c_D = 0$, $C_2(T, N)$ agrees with (8.17).

Similarly, when tasks are backed up at time T or the Nth completion of services, whichever occurs last,

$$C_2(T, N) = \frac{\begin{array}{c} c_F - (c_F - c_T)\overline{F}(T)G^{(N)}(T) - (c_F - c_N) \int_T^\infty \overline{F}(t)\mathrm{d}G^{(N)}(t) \\ + c_D[\mu - \int_T^\infty tG^{(N)}(t)\mathrm{d}F(t)] \end{array}}{\mu - \int_T^\infty G^{(N)}(t)\overline{F}(t)\mathrm{d}t}. \tag{8.90}$$

When $c_D = 0$, $C_2(T, N)$ agrees with (8.84).

8.3 Other Maintenance Models

8.3.1 Parallel System

Consider a parallel redundant system that consists of N identical units and fails when all units have failed. It was shown by graph that the system can operate for a specified mean time by either changing the replacement time or increasing the number of units [2]. The problem of determining the optimum number of units and the replacement number of failed units before system failure were discussed. The known results for these problems have been summarized in Chap. 2.

Suppose that each unit has an identical failure distribution $F(t)$ with a finite mean μ. A parallel system with N units ($N = 1, 2, \dots$) is replaced at time T or at system failure, whichever occurs first [34]. Then, because the system has a failure distribution $F(t)^N$, the mean time to replacement is

$$T[1 - F(T)^N] + \int_0^T t \, dF(t)^N = \int_0^T [1 - F(t)^N] \, dt. \qquad (8.91)$$

Let c_0 be an acquisition cost for one unit and c_F be an additional replacement cost for a failed system. Then, from (8.91), the expected cost rate is

$$C(T, N) = \frac{N c_0 + c_F F(T)^N}{\int_0^T [1 - F(t)^N] dt}. \qquad (8.92)$$

When $N = 1$, $C(T, 1)$, agrees with the expected cost rate for the standard age replacement [1, p. 72].

We find both optimum T^* and N^* that minimizes $C(T, N)$ in (8.92). From the inequalities $C(T, N+1) - C(T, N) \geq 0$,

$$L(N; T) \geq \frac{c_F}{c_0} \qquad (N = 1, 2, \dots), \qquad (8.93)$$

where

$$L(N; T) \equiv \frac{(N + 1) \int_0^T [1 - F(t)^N] dt - N \int_0^T [1 - F(t)^{N+1}] dt}{F(T)^N \int_0^T [1 - F(t)^{N+1}] dt - F(T)^{N+1} \int_0^T [1 - F(t)^N] dt}.$$

Because

$$(N + 1) \int_0^T [1 - F(t)^N] \, dt - N \int_0^T [1 - F(t)^{N+1}] \, dt$$

$$= \int_0^T \overline{F}(t)[1 + F(t) + \cdots + F(t)^{N-1} - N F(t)^N] \, dt > 0,$$

and

$$\lim_{N \to \infty} \left\{ (N + 1) \int_0^T [1 - F(t)^N] \, dt - N \int_0^T [1 - F(t)^{N+1}] \, dt \right\} = T,$$

the numerator of $L(N; T)$ is positive and tends to T as $N \to \infty$, where $\overline{F}(t) \equiv 1 - F(t)$. Similarly, the denominator is

$$F(T)^N \int_0^T \{\overline{F}(T) + F(t)^N [F(T) - F(t)]\} \, dt > 0,$$

that tends to 0 as $N \to \infty$ for $T > 0$. Thus, $\lim_{N \to \infty} L(N; T) = \infty$ for any $T > 0$. In addition, from the definition of $L(N; T)$,

$$L(N + 1; T) - L(N; T)$$

$$= A \left\{ [-(N + 2) F(T) + (N + 1) F(T)^2] \int_0^T [1 - F(t)^N] \, dt \right.$$

$$+ [N + 2 - N F(T)^2] \int_0^T [1 - F(t)^{N+1}] \, dt$$

$$\left. + [-(N + 1) + N F(T)] \int_0^T [1 - F(t)^{N+2}] \, dt \right\}$$

$$= A \left\{ T[1 - F(T)^2] + \int_0^T F(t)^N [F(T) - F(t)] \, [N\overline{F}(t)\overline{F}(T) + \overline{F}(t) + \overline{F}(T)] \, dt \right\}$$

$$> 0,$$

where

$$A \equiv \frac{\int_0^T [1 - F(t)^{N+1}] dt}{F(T)^{N+1} \left\{ \int_0^T [1 - F(t)^{N+2}] dt - F(T) \int_0^T [1 - F(t)^{N+1}] \, dt \right\}} \cdot$$
$$\times \left\{ \int_0^T [1 - F(t)^{N+1}] \, dt - F(T) \int_0^T [1 - F(t)^N] dt \right\}$$

Therefore, there exists a unique minimum N^* $(1 \le N^* < \infty)$ that satisfies (8.93) for any $T > 0$.

Next, assume that the unit has the failure rate $h(t)$. Then, differentiating $C(T, N)$ in (8.92) with respect to T and setting it equal to zero,

$$q(T; N) \int_0^T [1 - F(t)^N] \, dt - F(T)^N = \frac{Nc_0}{c_F}, \qquad (8.94)$$

where

$$q(T; N) \equiv h(T) \frac{N[F(T)^{N-1} - F(T)^N]}{1 - F(T)^N}.$$

If $h(t)$ increases strictly to ∞, then it is easily proved that $q(T; N)$ also increases strictly to ∞. Thus, the left-hand side of (8.94) also increases strictly from 0 to ∞, and hence, there exists a finite and unique T^* $(0 < T^* < \infty)$ that satisfies (8.94) for any $N \ge 1$. In this case, the resulting cost rate is

$$C(T^*; N) = c_F q(T^*; N). \qquad (8.95)$$

When $h(\infty) \equiv \lim_{t \to \infty} h(t) < \infty$, $\lim_{T \to \infty} q(T; N) = h(\infty)$, and hence, the left-hand side of (8.94) tends to

$$h(\infty) \int_0^\infty [1 - F(t)^N] \, dt - 1 \qquad \text{as } T \to \infty.$$

Thus, if $h(\infty) > (Nc_0 + c_F)/\{c_F \int_0^\infty [1 - F(t)^N] dt\}$, then a finite T^* to satisfy (8.94) exists uniquely.

From the above results, we have to solve the equations with two variables in computing the optimum T^* and N^*. We can specify the computing procedure for obtaining T^* and N^* when $h(t)$ increases strictly to ∞:

1. Set $N_0 = 1$ and compute T_1 to satisfy (8.94).
2. Set $T = T_1$ and compute N_1 to satisfy satisfy (8.93).
3. Set $N = N_1$ and compute T_2 to satisfy (8.94).
4. Continue until $N_k = N_{k+1}$ $(k = 0, 1, 2, \dots)$.

Table 8.4. Optimum number N^* and time T^*, expected cost rate $C(T^*, N^*)/c_0$, and N^* and $C(\infty, N^*)/c_0$ when $F(t) = 1 - \exp(-t^2)$

c_F/c_0	N^*	T^*	$C(T^*, N^*)/c_0$	N^*	$C(\infty, N^*)/c_0$
10	1	0.32	6.38	3	10.08
20	1	0.23	9.00	5	17.10
30	1	0.18	10.98	7	23.59
40	1	0.16	12.72	9	29.79
50	1	0.14	14.20	10	35.81
100	2	0.30	9.16	17	64.08
200	2	0.25	11.04	30	116.43
300	2	0.22	12.07	41	166.07
500	2	0.20	14.03	61	261.11
1000	2	0.16	16.51	108	486.26

Example 8.4. The computing procedure is convenient and rapid because it is hardly necessary to provide maintained systems with more redundancy. Table 8.4 presents the optimum T^* and N^* for c_F/c_0 when $F(t) = 1 - \exp(-t^2)$. This indicates that the optimum number N^* and the expected cost rate $C(T^*, N^*)$ are relatively smaller compared with those of the system with no planned replacement time, *i.e.*, $T = \infty$. It is of interest that the expected cost rate $C(0.30, 2)$ becomes lower than $C(0.14, 1)$ even if the cost c_F/c_0 becomes larger from 50 to 100. ■

8.3.2 Inspection Policies

(1) Periodic Inspection

The unit is checked at periodic times kT $(k = 1, 2, \ldots, N)$: Any failure is detected at the next check time, and the unit is replaced immediately. The optimum inspection policies for a finite time span have been discussed in Sects. 3.1.1 and 4.2. It is assumed that a prespecified number of warranty is N, *i.e.*, the system is replaced at time NT. Any check and replacement times are negligible. This is applied to a storage system that can be made only at a specified finite number of inspections [159]. For example, missiles are composed of various kinds of mechanical, electric and electronic parts, and some parts have a short life time because they have to generate high power in a very short operating time. Such parts should be exchanged after the total times of inspections have exceeded a prespecified time of quality warranty.

Let c_I be the cost for one check, c_D be the cost per unit of time for the time elapsed between a failure and its detection at the next check time, and

c_R be the replacement cost at time NT or at failure. Then, the expected cost when the unit is replaced because of its failure at time kT $(k = 1, 2, \ldots, N)$ is, from (4.6),

$$\sum_{k=1}^{N} \int_{(k-1)T}^{kT} [kc_I + (kT - t)c_D + c_R]\, dF(t), \tag{8.96}$$

and when it is replaced without failure at time NT,

$$\overline{F}(NT)(Nc_I + c_R). \tag{8.97}$$

Thus, the total expected cost until replacement is, from (8.96) and (8.97),

$$\sum_{k=1}^{N} \int_{(k-1)T}^{kT} [kc_I + (kT - t)c_D + c_R]\, dF(t) + \overline{F}(NT)(Nc_I + c_R)$$

$$= (c_I + c_D T) \sum_{k=0}^{N-1} \overline{F}(kT) - c_D \int_{0}^{NT} \overline{F}(t)\, dt + c_R. \tag{8.98}$$

Furthermore, the mean time to replacement is

$$\sum_{k=1}^{N} \int_{(k-1)T}^{kT} kT dF(t) + NT\overline{F}(NT) = T \sum_{k=0}^{N-1} \overline{F}(kT). \tag{8.99}$$

Therefore, the expected cost rate is, from (8.98) and (8.99),

$$C_1(T, N) = \frac{c_I \sum_{k=0}^{N-1} \overline{F}(kT) - c_D \int_{0}^{NT} \overline{F}(t)dt + c_R}{T \sum_{k=0}^{N-1} \overline{F}(kT)} + c_D. \tag{8.100}$$

It can be clearly seen that $\lim_{T\to 0} C_1(T, N) = \infty$ and $\lim_{T\to\infty} C_1(T, N) = c_D$. Thus, there exists a positive T^* $(0 < T^* \le \infty)$ that minimizes $C_1(T, N)$ for a specified $N \ge 1$.

When the failure time of the unit is exponential, i.e., $\overline{F}(t) = 1 - e^{-\lambda t}$, the expected cost rate is

$$C_1(T, N) = \frac{c_I}{T} + c_D - \frac{1}{\lambda T}(1 - e^{-\lambda T})\left(c_D - \frac{\lambda c_R}{1 - e^{-N\lambda T}}\right). \tag{8.101}$$

We investigate the properties of an optimum T^* that minimizes $C_1(T, N)$. Differentiating $C_1(T, N)$ with respect to T and setting it equal to zero,

$$\left(\frac{c_D}{\lambda} - \frac{c_R}{1 - e^{-N\lambda T}}\right)[1 - (1 + \lambda T)e^{-\lambda T}] - \frac{c_R N \lambda T e^{-N\lambda T}(1 - e^{-\lambda T})}{(1 - e^{-N\lambda T})^2} = c_I. \tag{8.102}$$

Denoting the left-hand side of (8.101) by $Q_N(T)$,

$$\lim_{T \to 0} Q_N(T) = -\frac{c_R}{N}, \qquad \lim_{T \to \infty} Q_N(T) = \frac{c_D}{\lambda} - c_R.$$

Furthermore, from (8.102),

$$Q_{N+1}(T) - Q_N(T) = c_R(1 - e^{-\lambda T})e^{-N\lambda T} \left\{ \frac{1 - (1 + \lambda T)e^{-\lambda T}}{(1 - e^{-N\lambda T})(1 - e^{-(N+1)\lambda T})} \right.$$

$$\left. + \lambda T \left[\frac{N}{(1 - e^{-N\lambda T})^2} - \frac{(N+1)e^{-\lambda T}}{(1 - e^{-(N+1)\lambda T})^2} \right] \right\}.$$

The first term in the bracket is clearly positive. The second term is

$$\frac{N}{(1 - e^{-N\lambda T})^2} - \frac{(N+1)e^{-\lambda T}}{(1 - e^{-(N+1)\lambda T})^2}$$

$$= \frac{N(1 - e^{-(N+1)\lambda T})^2 - (N+1)e^{-\lambda T}(1 - e^{-N\lambda T})^2}{(1 - e^{-N\lambda T})^2(1 - e^{-(N+1)\lambda T})^2}.$$

The numerator of the above right-hand side is

$$N(1 - e^{-(N+1)\lambda T})^2 - (N+1)e^{-\lambda T}(1 - e^{-N\lambda T})^2$$

$$= e^{-\lambda T}[N(e^{\lambda T} - 1)(1 - e^{-(2N+1)\lambda T}) - (1 - e^{-N\lambda T})^2] > 0.$$

Thus, $Q_N(T)$ also increases strictly with N.

From the above results, there exists a finite T^* $(0 < T^* < \infty)$ that satisfies (8.102) for $c_D/\lambda > c_I + c_R$. In addition, because $Q_N(T)$ increases with N, an optimum T^* decreases with N. When $N = 1$, from (8.102)

$$1 - (1 + \lambda T)e^{-\lambda T} = \frac{\lambda(c_I + c_R)}{c_D}, \qquad (8.103)$$

and when $N = \infty$,

$$1 - (1 + \lambda T)e^{-\lambda T} = \frac{\lambda c_I}{c_D - \lambda c_R}. \qquad (8.104)$$

Thus, $T_\infty^* \leq T^* < T_1^*$, where T_1^* and T_∞^* are the respective solutions of (8.103) and (8.104). The condition of $c_D/\lambda > c_I + c_R$ means that the total cost for the mean life of the system is greater than the summation of check and replacement costs. This would be realistic in actual fields.

Example 8.5. We compute the optimum time T^* that satisfies (8.102) for a specified number N. Table 8.5 presents the optimum T^* and the resulting cost rate $C_1(T^*, N)$ in (8.101) for $\lambda = 1.0 \times 10^{-3}$, 1.1×10^{-3}, and 1.2×10^{-3}, and $N = 1, 2, \ldots,$ and 10 when $c_I = 10$, $c_D = 1$, and $c_R = 100$. This indicates that T^* decreases with both λ and N, and $C_1(T^*, N)$ increases with λ and decreases with N. ■

Table 8.5. Optimum time T^* and expected cost rate $C_1(T^*, N)$ when $c_I = 10$, $c_D = 1$, and $c_R = 100$

N	$\lambda = 1.0 \times 10^{-3}$		$\lambda = 1.1 \times 10^{-3}$		$\lambda = 1.2 \times 10^{-3}$	
	T^*	$C_1(T^*, N)$	T^*	$C_1(T^*, N)$	T^*	$C_1(T^*, N)$
1	564	2.03	543	2.11	526	2.18
2	396	1.71	380	1.80	367	1.88
3	328	1.54	314	1.63	303	1.71
4	289	1.43	277	1.52	267	1.60
5	264	1.36	253	1.44	243	1.53
6	246	1.30	236	1.39	226	1.48
7	233	1.26	223	1.35	214	1.43
8	222	1.23	212	1.32	204	1.40
9	214	1.20	204	1.29	196	1.38
10	207	1.18	197	1.27	189	1.36

(2) Storage System

We consider a system in storage that is required to achieve a higher reliability than a prespecified level q $(0 < q < 1)$ [1, p. 216, 19]. To hold the reliability, the system is checked and is maintained at periodic times NT $(N = 1, 2, \ldots)$, and is replaced or overhauled if the reliability becomes equal to or lower than q. The total checking number N^* and the $N^*T + t_0$ until replacement are derived when the system reliability is just equal to q. Using them, the expected cost rate $C(T)$ until replacement is obtained, and an optimum checking time T^* that minimizes it is computed numerically. Two extended models were considered where the system is also replaced at time $(N + 1)T$ [160] and may be degraded at each checking time [161].

The system consists of unit 1 and unit 2, where the failure time of unit i has a cumulative hazard function $H_i(t)$ $(i = 1, 2)$. When the system is checked at periodic times NT $(N = 1, 2, \ldots)$, unit 1 is maintained and is like new after every check, and unit 2 is not done, i.e., its hazard rate remains unchanged by any inspection. In addition, it is assumed that any times required for check and maintenance are negligible. From such assumptions, the reliability function $R(t)$ of the system with no inspection is

$$R(t) = e^{-H_1(t) - H_2(t)}. \tag{8.105}$$

If the system is checked and maintained at time t, the reliability just after the check is

$$R(t_{+0}) = e^{-H_2(t)}. \tag{8.106}$$

Thus, the reliabilities just before and after the Nth check are, respectively,

$$R(NT_{-0}) = e^{-H_1(NT) - H_2(NT)}, \qquad R(NT_{+0}) = e^{-H_2(NT)}. \qquad (8.107)$$

Next, suppose that the replacement or overhaul is done if the system reliability is equal to or lower than q. Then , if

$$e^{-H_1(T) - H_2(NT)} > q \geq e^{-H_1(T) - H_2[(N+1)T]}, \qquad (8.108)$$

the time to replacement is $NT + t_0$, where t_0 $(0 < t_0 \leq T)$ satisfies

$$e^{-H_1(t_0) - H_2(NT + t_0)} = q. \qquad (8.109)$$

This shows that the reliability is greater than q just before the Nth check and is equal to q at time $NT + t_0$.

Let c_I and c_R be the respective costs for check and replacement. Then, the expected cost rate until replacement is given by

$$C_2(T, N) = \frac{Nc_I + c_R}{NT + t_0}. \qquad (8.110)$$

Example 8.6. When the failure time of units has an exponential distribution, i.e., $H_i(t) = \lambda_i t$ $(i = 1, 2)$, (8.108) becomes

$$\frac{1}{Na + 1} \log \frac{1}{q} \leq \lambda T < \frac{1}{(N-1)a + 1} \log \frac{1}{q}, \qquad (8.111)$$

where $\lambda \equiv \lambda_1 + \lambda_2$ and $a \equiv H_2(T)/[H_1(T) + H_2(T)] = \lambda_2/\lambda$ $(0 < a < 1)$ that represents an efficiency of inspection.

When an inspection time T is given, an inspection number N^* that satisfies (8.111) is determined. In this case, (8.109) is

$$N^* \lambda_2 T + \lambda t_0 = \log \frac{1}{q}. \qquad (8.112)$$

Thus, the total time to replacement is

$$N^* T + t_0 = N^*(1 - a)T + \frac{1}{\lambda} \log \frac{1}{q}, \qquad (8.113)$$

and the expected cost rate is

$$C_2(T, N^*) = \frac{N^* c_I + c_R}{N^*(1 - a)T + (1/\lambda) \log(1/q)}. \qquad (8.114)$$

Therefore, when an inspection time T is given, we compute N^* from (8.111) and $N^*T + t_0$ from (8.113). Substituting these values in (8.114), we get $C_2(T, N^*)$. Changing T from 0 to $\log(1/q)/[\lambda(1 - a)]$, because λT is less than $\log(1/q)/(1 - a)$ from (8.111), we can compute an optimum T^* that minimizes $C_2(T, N^*)$. When $\lambda T \geq \log(1/q)/(1 - a)$, $N^* = 0$ and

Table 8.6. Optimum number N^* and time to replacement $\lambda(N^*T+t_0)$ when $a = 0.1$ and $q = 0.8$

λT	N^*	$\lambda(N^*T + t_0)$
$[0.223, \infty)$	0	$[0.223, \infty)$
$[0.203, 0.223)$	1	$[0.406, 0.424)$
$[0.186, 0.203)$	2	$[0.558, 0.588)$
$[0.172, 0.186)$	3	$[0.687, 0.725)$
$[0.159, 0.172)$	4	$[0.797, 0.841)$
$[0.149, 0.159)$	5	$[0.893, 0.940)$
$[0.139, 0.149)$	6	$[0.976, 1.026)$
$[0.131, 0.139)$	7	$[1.050, 1.102)$
$[0.124, 0.131)$	8	$[1.116, 1.168)$
$[0.117, 0.124)$	9	$[1.174, 1.227)$
$[0.112, 0.117)$	10	$[1.227, 1.280)$

Table 8.7. Optimum number N^*, time λT^*, time to replacement $\lambda(N^*T^* + t_0)$, and expected cost rate $C_2(T^*, N^*)/\lambda$

c_R/c_I	a	q	N^*	λT^*	$\lambda(N^*T^* + t_0)$	$C_2(T^*, N^*)/\lambda$
10	0.1	0.8	8	0.131	1.168	15.41
50	0.1	0.8	19	0.080	1.586	43.51
10	0.5	0.8	2	0.149	0.372	32.27
10	0.1	0.9	7	0.062	0.552	32.63

$$C_2(T, 0) = \frac{c_R}{t_0} = \frac{\lambda c_R}{\log(1/q)}. \tag{8.115}$$

Table 8.6 represents the optimum number N^* and the total time $\lambda(N^*T + t_0)$ to replacement for λT when $a = 0.1$ and $q = 0.8$. For example, when λT increases from 0.203 to 0.223, $N^* = 1$ and $\lambda(N^*T + t_0)$ increases from 0.406 to 0.424. In accordance with the decrease in λT, both N^* and $\lambda(N^*T + t_0)$ increase as shown in (8.111) and (8.113), respectively.

Table 8.7 represents the optimum number N^* and time T^* that minimize $C_2(T, N)$ for c_R/c_I, a, and q, the resulting total time $\lambda(N^*T^* + t_0)$, and the expected cost rate $C_2(T^*, N^*)/\lambda$ for $c_I = 1$. These indicate that λT^* increases and $\lambda(N^*T^* + t_0)$ decreases when c_I/c_R and a increase, and both λT^* and $\lambda(N^*T^* + t_0)$ decrease when q increases. ■

9

System Complexity and Entropy Models

The science of complexity has been developed widely in many fields such as physics, economics, and mathematics [162, 163]. In modern information societies, both hardware and software become more complex with increasing requirements of high quality and performance. It is well-known that the reliability of large-scale systems becomes lower than our expectation, owing to the complex of communication networks and the increase of hardware such as fault detection and switchover equipment [164, 165]. It is important to do further studies on system complexity in the field of reliability theory.

In this chapter, we define the complexity of redundant systems and calculate the reliabilities of typical systems with complexity. Several appropriate examples to understand these results easily are given. In Sect. 9.1, we define the complexity of redundant systems as the number of paths [166]. Two reliability functions of complexity are introduced, and reliabilities of standard redundant systems are calculated. An optimum number of units that maximizes the reliability of a parallel system is computed numerically.

As another measure of complexity, it would be a natural consequence to introduce the concept of entropy that represents the vagueness of incomplete information [167–170]. The notion of entropy has already been applied to reliability problems. For example, there have been many papers that treat the estimation of probability distributions based on the maximum entropy principle [171, 172]. Furthermore, it was shown [173] that the optimum safety monitoring system with n sensors composes a k-out-of-n system by using the conditional entropy [174]. Many measures of software complexity for computer systems were suggested to quantify it [175–178]. In addition, the complexity measure for emergency operating procedures was developed based on entropy measures in software engineering [179].

In Sect. 9.2, we define the complexity of redundant systems as a logarithmic function of the number of paths by using the concept of entropy [180]. Furthermore, we introduce a reliability function of complexity and calculate the reliabilities of series and parallel systems. As one typical redundant system, we deal with a majority decision system and determine numerically an optimum system that maximizes its reliability. Finally, we also propose the complexity of network systems.

However, there exist many theoretical and practical problems on system complexity that have to be solved from now on. We present briefly further studies concerned with complexity [180]:

(1) Show how to define the complexity of more complex systems where the number of paths and entropy cannot be computed.
(2) Show how to compute the reliability of a system with complexity when two reliabilities of an original system and its complexity are interactive with each other.
(3) Define the complexity of a network system whose entropy cannot be computed.
(4) Estimate parameters of the reliability functions of an original system and its complexity from actual data.

The entropy model has been proposed as applied models of information theory and adapted practically to some actual problems in several fields of operations research [170]. In Sect. 9.3, we attempt to apply the entropy model to maintenance problems in reliability theory. When two replacement costs after and before failure for an age replacement policy are given, two replacement rates after and before failure are derived by using the entropy model. Furthermore, these results are compared numerically with optimum age replacement times [1, p. 76]. It is shown that this can be applied to other maintenance models. It would be necessary to verify fully that the entropy model can be applied properly to actual maintenance models.

9.1 System Complexity

9.1.1 Definition of Complexity

A system is usually more complicated as the number of units or modules in it increases, so that it might be reasonable roughly to define the system complexity as the number of units or modules. However, if most units of a system are composed in series, its reliability decreases as the number increases, and it is not so much complicated from the viewpoint of reliability [166]. When the number of paths of a system is known, it would be natural to define the complexity as the number of paths rather than the number of units.

Consider a system with two terminals and n units [2]: The performance of each unit is represented by an indicator x_i ($i = 1, 2, \ldots, n$) that takes 1 if it operates and 0 if it fails. The performance of the system depends on the performance of each unit and is represented by $\varphi(x)$ that also takes 1 if it operates and 0 if it fails, where $x = (x_1, x_2, \ldots, x_n)$. Then, we denote a partition A of the set of units as $A \equiv \{i : x_i = 1\}$. If $\varphi(x) = 1$ and $\varphi(y) = 0$ for any $y \leq x$ but $\neq x$, then A is a *path* of the system.

Fig. 9.1. Series system with n units

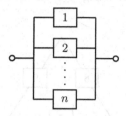

Fig. 9.2. Parallel system with n units

Suppose that we can count the number of paths of a system with two terminals and define the complexity as the number P_a of paths [166]:

(1) *Series system.* The number of paths of a series system with n units in Fig. 9.1 is 1. The complexity of the system is $P_a = 1$, independent of n.

(2) *Parallel system.* The number of paths of a parallel system with n units in Fig. 9.2 is n. The complexity is $P_a = n$.

(3) *k-out-of-n system.* Consider a k-out-of-n system that can operate if at least k units operate [2, p. 216]. The complexity is $P_a = \binom{n}{k}$ because the number of paths is $\binom{n}{k}$. In particular, when $k = 2$ and $n = 3$, *i.e.*, the system consists of a 2-out-of-3 system, $P_a = 3$.

(4) *Standby system.* The number of paths of a standby system where one unit is operating and $n - 1$ units are in standby is assumed to be equal to that of a parallel system. The complexity is $P_a = n$.

(5) *Bridge system.* The number of paths of a bridge system with 5 units in Fig. 9.3 is 4 [2], and hence, its complexity is $P_a = 4$.

(6) *Network system.* The number of paths of a network system with 9 units in Fig. 9.4 is 10 [2], and hence, its complexity is $P_a = 10$.

Next, suppose that a system is composed of several modules each of which is composed of several units and the complexity of module M_j is $P_a(j)$ ($j = 1, 2, \dots$).

(7) *Series system with m modules.* The number of paths is $\prod_{j=1}^{m} P_a(j)$. The system complexity is $P_a = \prod_{j=1}^{m} P_a(j)$, *i.e.*, it is given by the product of

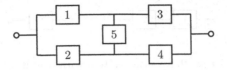

Fig. 9.3. Bridge system with five units

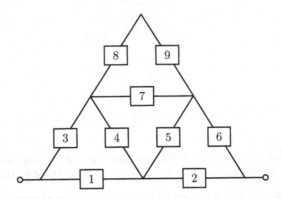

Fig. 9.4. Network system with nine units

complexity of each module. For example, the complexity of a series-parallel system in Fig. 9.5 is $P_a = n^m$.

(8) *Parallel system with m modules.* The number of paths is $\sum_{j=1}^{m} P_a(j)$. The complexity is $P_a = \sum_{j=1}^{m} P_a(j)$, *i.e.*, it is given by the summation of complexities of each module. The complexities of a parallel-series system in Fig. 9.6 is $P_a = m$, independent of n.

9.1.2 Reliability of Complexity

We specify reliability functions of complexity that are functions of the number of paths n and decrease from 1 to 0. Typical discrete functions chosen are geometric and discrete Weibull distributions [1, p. 17]. We use the following two reliability functions of complexity when $P_a = n$:

$$R_c(n) \equiv e^{-\alpha(n-1)} \equiv q^{n-1} \qquad (n = 1, 2, \dots), \qquad (9.1)$$

where $q \equiv e^{-\alpha}$ ($0 < \alpha < \infty$, $0 < q < 1$), and

$$R_c(n) \equiv e^{-\alpha(n-1)^\beta} \equiv q^{(n-1)^\beta} \qquad (n = 1, 2, \dots) \qquad (9.2)$$

for $\beta > 0$. When $\beta = 1$, $R_c(n)$ is equal to that of (9.1). Note that the reliability functions in (9.1) and (9.2) correspond to geometric and discrete Weibull distributions [1, p. 17], respectively.

Next, consider a system with complexity and compute its reliability that combines those of an original system and its complexity. Those reliabilities would be interactive on an original system and its complexity, however, it would be difficult to define such reliability theoretically. When the reliabilities of the system and the complexity P_a are given by R and $R_c(P_a)$, respectively, we define formally the reliability of the system with complexity as $R_s \equiv R_c(P_a) \times R$.

Assume that each unit has an identical reliability function R_0 and $R_c(P_a) = q^{P_a-1}$ from (9.1).

(9) *Series system.* The reliability of a series system in Fig. 9.1 is $R = R_0^n$ and $R_c(P_a) = 1$ from **(1)**. Thus, $R_s = R_0^n$, *i.e.*, we need not consider the complexity of any series systems.

(10) *Parallel system.* The reliability of a parallel system is $R = 1-(1-R_0)^n$ and $R_c(P_a) = q^{n-1}$ from **(2)**. Thus,

$$R_s = q^{n-1}[1 - (1 - R_0)^n]. \tag{9.3}$$

(11) *k-out-of-n system.* The reliability of a k-out-of-n system is

$$R = \sum_{j=k}^{n} \binom{n}{j} R_0^j (1 - R_0)^{n-j},$$

and $R_c(P_a) = q^{P_a-1}$, where $P_a = \binom{n}{k}$ from **(3)**. Therefore,

$$R_s = q^{P_a-1} \sum_{j=k}^{n} \binom{n}{j} R_0^j (1 - R_0)^{n-j}. \tag{9.4}$$

It might be better to define the reliability of a k-out-of-n system with complexity as

$$R_s = \sum_{j=k}^{n} q^{\binom{n}{j}-1} \binom{n}{j} R_0^j (1 - R_0)^{n-j}. \tag{9.5}$$

(12) *Series-parallel system.* The reliability of a series-parallel system in Fig. 9.5 is $R = [1 - (1 - R_0)^n]^m$ and $R_c(P_a) = q^{n^m-1}$ from **(7)**. Therefore,

$$R_s = q^{n^m-1}[1 - (1 - R_0)^n]^m. \tag{9.6}$$

In general, the reliability of a parallel system increases as the number of units increases. However, it would not be necessarily so if we consider a general idea of complexity because the reliability might decrease as the complexity

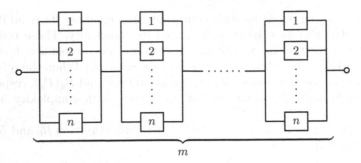

Fig. 9.5. Series-parallel system

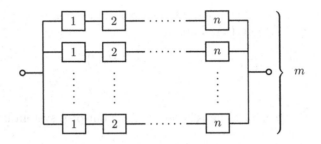

Fig. 9.6. Parallel-series system

increases. Finally, we discuss optimum numbers of units that maximize the reliability R_s and calculate them numerically.

(13) *Parallel system.* We calculate an optimum number n^* that maximizes

$$R_s(n) = q^{n-1}[1 - (1 - R_0)^n] \qquad (n = 1, 2, \ldots) \qquad (9.7)$$

for $0 < q < 1$. It is clearly seen that $R_s(1) = R_0$ and $\lim_{n \to \infty} R_s(n) = 0$. From the inequality $R_s(n+1) - R_s(n) \leq 0$,

$$(1 - R_0)^n \leq \frac{1 - q}{1 - q + qR_0}. \qquad (9.8)$$

Thus, an optimum n^* $(1 \leq n^* < \infty)$ is given by a unique minimum integer that satisfies (9.8). When $q(2 - R_0) \leq 1$, $n^* = 1$, *i.e.*, we should compose no redundant system. Table 9.1 presents the optimum number n^* for R_0 and α when $q = e^{-\alpha}$. This indicates that values of n^* decrease with both R_0 and α and become very small when the reliability of units is high and the reliability of complexity is low.

(14) *Majority decision system.* A majority decision system corresponds to an $(n+1)$-out-of-$(2n+1)$ system, and hence, its reliability is, from **(11)**,

Table 9.1 Optimum number n^* for an n-unit parallel system

R_0	α							
	10^{-1}	10^{-2}	10^{-3}	10^{-4}	10^{-5}	10^{-6}	10^{-7}	10^{-8}
$1 - 10^{-1}$	1	2	3	4	5	6	7	8
$1 - 10^{-2}$	1	1	2	2	3	3	4	4
$1 - 10^{-3}$	1	1	1	2	2	2	3	3

Table 9.2 Optimum number n^* for an $(n+1)$-out-of-$(2n+1)$ system

R_0	α							
	10^{-1}	10^{-2}	10^{-3}	10^{-4}	10^{-5}	10^{-6}	10^{-7}	10^{-8}
$1 - 10^{-1}$	1	1	2	3	4	5	6	7
$1 - 10^{-2}$	1	1	1	1	2	2	3	3
$1 - 10^{-3}$	1	1	1	1	1	1	2	2

$$R_s(n) = q^{P_a - 1} \sum_{j=n+1}^{2n+1} \binom{2n+1}{j} R_0^j (1 - R_0)^{2n+1-j} \qquad (n = 1, 2, \ldots), \quad (9.9)$$

where $P_a = \binom{2n+1}{n}$. Table 9.2 presents the optimum number n^* that maximizes $R_s(n)$ for R_0 and α. For example, when $R_0 = 0.9$ and $\alpha = 0.001$, *i.e.*, $q = 0.009$, $n^* = 2$. This indicates that a 3-out-of-5 system is the best under such conditions.

(15) *Series-parallel system.* The reliability of a series-parallel system with complexity is, from **(12)**,

$$R_s(n, m) = q^{n^m - 1}[1 - (1 - R_0)^n]^m \qquad (n, m = 1, 2, \ldots). \tag{9.10}$$

Table 9.3 presents the reliability $R_s(n, m)$ for m and n when $\alpha = 10^{-3}$ and $R_0 = 1 - 10^{-1}$, $1 - 10^{-2}$, and $1 - 10^{-3}$. It is clearly seen that the reliability decreases with m, however, it is maximum at $n = 3, 2, 1$ for $R_0 = 1 - 10^{-1}$, $1 - 10^{-2}$, and $1 - 10^{-3}$, respectively.

(16) *Multi-unit system and duplex system.* Consider the multi-unit system in Fig. 9.7 and the duplex system in Fig. 9.8, both of which have four identical units with reliability R_0. Then, the reliability of a multi-unit system with complexity is

$$R_s = q^3[1 - (1 - R_0)^2]^2,$$

because the number of paths is four, and the reliability of a duplex system is

Table 9.3. Reliability $R_s(n, m)$ of a series-parallel system

$R_0 = 0.9$

m	n				
	1	2	3	4	5
1	0.9000	0.9890	0.9970	0.9969	0.9960
2	0.8092	0.9772	0.9930	0.9928	0.9910
3	0.7275	0.9655	0.9891	0.9888	0.9861
4	0.6541	0.9539	0.9851	0.9847	0.9811

$R_0 = 0.99$

m	n				
	1	2	3	4	5
1	0.99000	0.99890	0.99800	0.99700	0.99601
2	0.97912	0.99681	0.99501	0.99302	0.99104
3	0.96836	0.99471	0.99203	0.98906	0.98610
4	0.95772	0.99263	0.98906	0.98511	0.98118

$R_0 = 0.999$

m	n				
	1	2	3	4	5
1	0.999000000	0.998999501	0.998001998	0.997004496	0.996007989
2	0.997003498	0.997002501	0.995012477	0.993024443	0.991040379
3	0.995010986	0.995009494	0.992031912	0.989060279	0.986097544
4	0.993022456	0.993020471	0.989060275	0.985111940	0.981179362

$$R_s = q[1 - (1 - R_0^2)^2].$$

In general, the reliability of a multi-unit system with no complexity, *i.e.*, $q = 1$, is higher than that of a duplex system. However, taking complexity into consideration, it is better than that of a duplex system only if $q > \sqrt{2 - R_0^2}/(2 - R_0)$. In particular, when $q = \sqrt{2 - R_0^2}/(2 - R_0)$, *i.e.*,

$$R_0 = \frac{2q^2 + \sqrt{2 - 2q^2}}{1 + q^2},$$

both reliabilities agree with each other. Table 9.4 indicates the values of q and α for R_0. For example, when $R_0 = 0.9$, if $q > 0.992$, *i.e.*, $\alpha < 8.3 \times 10^{-3}$, then a multi-unit system is better than a duplex system, and *vice versa*.

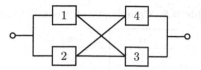

Fig. 9.7. Multi-unit system with four units

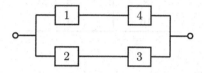

Fig. 9.8. Duplex system with four units

Table 9.4. Values of q and α when two reliabilities are equal

R_0	q	α
$1 - 10^{-1}$	0.992	8.3×10^{-3}
$1 - 10^{-2}$	0.9999	9.8×10^{-5}
$1 - 10^{-3}$	0.999999	9.0×10^{-7}

9.2 System Complexity Considering Entropy

We have defined system complexity as the number of paths of redundant systems with two terminals. However, as another measure of complexity, it would be natural to introduce the concept of entropy [180].

9.2.1 Definition of Complexity

We define the complexity of redundant systems as a logarithmic function to the base 2 of the number of paths, using the concept of entropy. Suppose that the number of paths of a system with two terminals is countable in the same way as that in Sect. 9.1. When the number of minimal paths is P_a, we define the system complexity as $P_e \equiv \log_2 P_a$. It is clearly seen that $P_e \leq P_a - 1$.

(17) *Series system.* The number of paths of a series system with n units in Fig. 9.1 is $P_a = 1$ from **(1)**. Thus, the complexity of the system is $P_e = \log_2 1 = 0$, independent of n.

(18) *Parallel system.* The number of paths of a parallel system with n units in Fig. 9.2 is $P_a = n$. Thus, the complexity of the system is $P_e = \log_2 n$. Table 9.5 presents the complexity of a parallel system for $n = 1, 2, 3, 4$,

Table 9.5. Complexity of a parallel system

n	$\log_2 n$
1	0
2	1
3	1.585
4	2
8	3
16	4

8, and 16. It is clearly seen that when the number of units is doubled, the complexity increases by 1 because the base of a logarithm is 2 in the same definition as that of entropy in information theory.

Next, suppose that each module is composed of several units and the number of paths and the complexity of module M_j are n_j and $P_e(j)$ $(j = 1, 2, \dots)$, respectively.

(19) *Series system with m modules.* Because the number of paths is $\prod_{j=1}^{m} n_j$ from **(7)**, the complexity is

$$P_e = \log_2 \left(\prod_{j=1}^{m} n_j \right) = \sum_{j=1}^{m} \log_2 n_j = \sum_{j=1}^{m} P_e(j). \tag{9.11}$$

The complexity of a series system is given by the summation of those of each module. This fact corresponds to the result that the failure rate of a series system is the total summation of those of each module. Thus, if a system can be divided into some modules in series even though it is complex, we can compute its complexity easily.

(20) *Parallel system with m modules.* Because the number of paths is $\sum_{j=1}^{m} n_j$ from **(8)**, the complexity is

$$P_e = \log_2 \left(\sum_{j=1}^{m} n_j \right). \tag{9.12}$$

In particular, when $n_j = n$, $P_e = \log_2 n + \log_2 m$, *i.e.*, the complexity is the summation of those of each module and parallel system with m units.

9.2.2 Reliability of Complexity

From a similar viewpoint and mathematical perspective to Sect. 9.1, we define the reliability function of complexity when $P_e = \log_2 n$:

$$R_e(n) \equiv e^{-\alpha P_e} = \exp(-\alpha \log_2 n) \qquad (n = 1, 2, \dots) \qquad (9.13)$$

for parameter $\alpha > 0$, that decreases with n from 1 to 0 and is higher than $R_c(n)$ in (9.1) for $n \geq 3$. The failure rate (hazard rate) of the reliability is

$$\frac{R_e(n) - R_e(n+1)}{R_e(n)} = 1 - \exp\{-\alpha[\log_2(n+1) - \log_2 n]\}, \qquad (9.14)$$

that decreases strictly from $1 - e^{-\alpha}$ to 0, that is, the complexity has the DFR (Decreasing Failure Rate) property [1, p. 6].

(21) *Series system with m modules.* The reliability of the complexity of a series system with m modules is, from (9.11) and (9.13),

$$R_e(P_e) = \exp\left(-\alpha \sum_{j=1}^{m} \log_2 n_j\right) = \prod_{j=1}^{m} R_e(n_j). \qquad (9.15)$$

The reliability of a series system with complexity is equivalent to the product of those of each module. This corresponds to the well-known result that the reliability of a series system is given by the product of those of each module.

(22) *Parallel system with m modules.* The reliability of the complexity of a parallel system with m modules is, from (9.12) and (9.13),

$$R_e(P_e) = \exp\left[-\alpha \log_2\left(\sum_{j=1}^{m} n_j\right)\right]. \qquad (9.16)$$

It is assumed that when the reliabilities of a system and its complexity are given by R and $R_e(P_e)$, respectively, we define the reliability of the system with complexity as $R_s = R_e(P_e) \times R$. Then, we compute numerically optimum numbers n^* that maximize the reliabilities of a parallel system and a majority decision system.

(23) *Parallel system.* The reliability of a parallel system with complexity is, from **(10)** and **(22)**,

$$R_s(n) = \exp(-\alpha \log_2 n)[1 - (1 - R_0)^n] \qquad (9.17)$$

for $0 < R_0 < 1$. It is clearly seen that $R_s(1) = R_0$ and $\lim_{n \to \infty} R_s(n) = 0$. Table 9.6 presents the optimum number n^* that maximizes $R_s(n)$ for R_0 and α. It is of interest that the optimum values n^* are not less than those in Table 9.1. This indicates that we should adopt a system with more units because its reliability R_e is equal to or higher than R_c in **(13)**.

(24) *Majority decision system.* Consider a majority decision system, *i.e.*, an $(n+1)$-out-of-$(2n+1)$ system. Because the complexity of the system is $\log_2 \binom{2n+1}{n}$ from **(3)**, its reliability is, from (9.13),

Table 9.6. Optimum number n^* for an n-unit parallel system

R_0	α							
	10^{-1}	10^{-2}	10^{-3}	10^{-4}	10^{-5}	10^{-6}	10^{-7}	10^{-8}
$1 - 10^{-1}$	1	3	4	5	6	7	8	9
$1 - 10^{-2}$	1	1	2	3	3	4	4	5
$1 - 10^{-3}$	1	1	1	2	2	3	3	3

Table 9.7. Optimum number n^* for a majority decision system

R_0	α							
	10^{-1}	10^{-2}	10^{-3}	10^{-4}	10^{-5}	10^{-6}	10^{-7}	10^{-8}
$1 - 10^{-1}$	1	2	3	6	8	10	12	14
$1 - 10^{-2}$	1	1	1	2	2	3	4	4
$1 - 10^{-3}$	1	1	1	1	1	2	2	2

$$R_e(n) = \exp\left[-\alpha \log_2 \binom{2n+1}{n}\right] \qquad (n = 1, 2, \ldots). \tag{9.18}$$

Thus, when the system consists of identical units with reliability R_0, the reliability of the system with complexity is, from **(11)**,

$$R_s(n) = \exp\left[-\alpha \log_2 \binom{2n+1}{n}\right] \sum_{j=n+1}^{2n+1} \binom{2n+1}{j} R_0^j (1 - R_0)^{2n+1-j}$$

$$(n = 1, 2, \ldots). \tag{9.19}$$

Table 9.7 presents the optimum number n^* that maximizes $R_s(n)$ for R_0 and α. This indicates that the optimum n^* are not less than those in Table 9.2 and become small for large α, *i.e.*, we should adopt a majority decision system with small units when the reliability of the complexity is low.

Finally, consider the complexity of a network system: The computational complexity of network reliability was summarized [181]. Four algorithms for computing network reliability with two terminals were compared [182]. On the other hand, it was proposed that a complexity measure is the total number of input-output paths in a network [183]. However, it would be meaningless to compute a network complexity as the number of paths by the method similar to redundant systems. We give two examples of network systems and their complexities.

(25) *Network systems.* We consider a network system with two terminals that consists of n networks in Fig. 9.9. When the relative frequency of usages

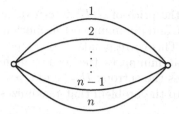

Fig. 9.9. Network system with two terminals

Fig. 9.10. Network system with four nodes

for network j $(j = 1, 2, \ldots, n)$ is estimated as P_j, where $\sum_{j=1}^{n} P_j = 1$, we define the complexity of the system as $P_e \equiv -\sum_{j=1}^{n} P_j \log_2 P_j$, using the definition of entropy. It is clearly seen that P_e is maximized when $P_j = 1/n$, and $P_e = \log_2 n$ that is equal to the complexity of a parallel system. For example, when the respective relative frequencies of usages for four networks are $P_1 = 1/2$, $P_2 = 1/4$, $P_3 = P_4 = 1/8$, the complexity is $P_e = 7/4$, and decreases by $2 - 7/4 = 1/4$, compared with the case of unknown frequencies of usages.

Next, we take up a network system with four nodes in Fig. 9.10. In this case, counting the number of networks that are connected with nodes, we may define the complexity and its reliability as $P_e = \log_2 6$ and $R_e(P_e) = \exp(-\alpha \log_2 6)$, respectively. If we could compute the entropy of network systems by any possible means, then it might be regarded as their complexity.

9.3 Entropy Models

We apply the entropy model [167–170] to maintenance models [1]. First, we introduce the following simple model [170]: Suppose that there are two brands A and B of an article on the market. A consumer usually buys either A or B at one's free will, taking into consideration their quality, price, facility, and one's past experience. However, it would be impossible to take a certain measure of free will from the public and be reasonable to judge that a consumer buys A or B with some unmethodical measure.

It is assumed that the price of $A(B)$ is $c_1(c_2)$, respectively. A consumer buys $A(B)$ with free will and minimum cost as much as possible with selection rate $p(q)$, respectively. This corresponds to one kind of optimum problem in operations research that minimizes the mean purchase cost $c_1 p + c_2 q$ and coincidentally maximizes the entropy $H = -p \log p - q \log q$. To simplify this problem, we change it to the problem that maximizes [170]

$$C(p, q) = \frac{-p \log p - q \log q}{c_1 p + c_2 q} \tag{9.20}$$

subject to $p + q = 1$, where the base 2 of the logarithm is omitted to simplify equations.

Denote that $c_1 : c_2 = l_1 : l_2$, where l_1 and l_2 are natural numbers and l_1/l_2 is written in the lowest term. For example, when $c_1 = 100$ and $c_2 = 1000$, $l_1 = 1$ and $l_2 = 10$. Then, using the method of undetermined multiplier λ of Lagrange, this becomes the problem that maximizes

$$F(p, q) = C(p, q) + \lambda(p + q - 1).$$

This problem is easily solved as follows: Differentiating $F(p, q)$ with respect to p and q and setting them equal to zero,

$$\log p = -\log e - \frac{l_1 H}{\bar{l}} + \lambda \bar{l}, \tag{9.21}$$

$$\log q = -\log e - \frac{l_2 H}{\bar{l}} + \lambda \bar{l}, \tag{9.22}$$

where $\bar{l} \equiv l_1 p + l_2 q$. Thus, $\lambda = \log e / \bar{l}$, using $H = -p \log p - q \log q$. Furthermore, from (9.21) and (9.22),

$$p = 2^{-l_1(H/\bar{l})}, \qquad q = 2^{-l_2(H/\bar{l})}. \tag{9.23}$$

Setting $W_0 \equiv 2^{H/\bar{l}}$, it follows that

$$W_0^{-l_1} + W_0^{-l_2} = 1, \tag{9.24}$$

because $p + q = 1$. Note that there exists always a positive and unique W_0 that satisfies (9.24) for any natural numbers l_1 and l_2. It is of interest that when $l_1 = 1$ and $l_2 = 2$, W_0 takes the golden ratio appeared in Sect. 2.2. Thus, the optimum p^* and q^* that maximize $C(p, q)$ in (9.20) are given by

$$p^* = W_0^{-l_1}, \qquad q^* = W_0^{-l_2}. \tag{9.25}$$

This is naturally generalized as follows [170]: The price of brand A_j is c_j and its selection rates are p_j $(j = 1, 2, \ldots, n)$, where $\sum_{j=1}^{n} p_j = 1$. When $c_1 : c_2 : \cdots : c_n = l_1 : l_2 : \cdots : l_n$, where l_j is a natural number and their rates are written in the lowest possible form, optimum p_j^* that maximize

$$C(p_1, p_2, \ldots, p_n) = \frac{-\sum_{j=1}^n p_j \log p_j}{\sum_{j=1}^n c_j p_j} \tag{9.26}$$

are given by

$$p_j^* = W_0^{-l_j} \qquad (j = 1, 2, \ldots, n), \tag{9.27}$$

where W_0 is a positive and unique solution of the equation $\sum_{j=1}^n W_0^{-l_j} = 1$.

If we would not be subject to restrictions on maintenance times, costs, and circumstances of units, then we could want to act freely. From such viewpoints, we can apply the above entropy model to some standard maintenance policies:

(26) *Age replacement model.* Suppose that an operating unit fails according to a failure distribution $F(t)$, where $\overline{F}(t) \equiv 1 - F(t)$. To prevent the failure, the unit is replaced before failure at time T $(0 < T \le \infty)$ or at failure, whichever occurs first [1, p. 69]. It is assumed that c_1 is the replacement cost for a failed unit and $c_2(< c_1)$ is the replacement cost for a non-failed unit at time T. We set $p \equiv F(T)$ and $q \equiv \overline{F}(T)$. The simplest method of age replacement is to balance the replacement at failure against that at non-failure, *i.e.*, $c_1 p = c_2 q$. In this case,

$$\widehat{p} = \frac{c_2}{c_1 + c_2}.$$

Next, suppose that the failure rate is $h(t) \equiv f(t)/\overline{F}(t)$, where $f(t)$ is a density function of $F(t)$. Then, the expected cost rate for the age replacement is [1, p. 72]

$$C(T) = \frac{c_1 F(T) + c_2 \overline{F}(T)}{\int_0^T \overline{F}(t)dt}, \tag{9.28}$$

and an optimum T^* to minimize $C(T)$ is given by a solution of the equation

$$h(T) \int_0^T \overline{F}(t)dt - F(T) = \frac{c_2}{c_1 - c_2}. \tag{9.29}$$

When the failure distribution is Weibull, *i.e.*, $\overline{F}(t) = \exp[-(\lambda t)^m]$, a finite T^* $(0 < T^* < \infty)$ that satisfies (9.29) for $m > 1$ exists uniquely.

We compare p^*, $\widehat{p}$, and $F(T^*)$ numerically when $\overline{F}(t) = \exp[-(\lambda t)^m]$. Table 9.8 presents p^*, $\widehat{p}$, and $F(T^*)$ for $c_1/c_2 = 2, 4, 6$, and 10 and $m = 1.6$, 2.0, 2.4, and 3.0. Note that p^* and $\widehat{p}$ exist apart from any failure distributions. This indicates that $p^* > \widehat{p}$ for any c_1/c_2, and p^* are closely $F(T^*) \times 100$ around $c_1/c_2 = 5$ and $m = 2.0$. It would be necessary to verify whether or not the values of p^* are proper, compared with actual maintenance data.

When failures occur very rarely, it might often be difficult to estimate the replacement cost c_1 at failure because this includes all costs resulting from a failure and its risk. From $\widehat{p} = c_2/(c_1 + c_2)$ and (9.25), we have the relations

$$\frac{1 - \widehat{p}}{\widehat{p}} = \frac{c_1}{c_2}, \qquad \frac{\log p^*}{\log(1 - p^*)} = \frac{c_1}{c_2}.$$

Table 9.8. Optimum p^*, $\widehat{p}$, and $F(T^*)$ when $\overline{F}(t) = \exp[-(\lambda t)^m]$

c_1/c_2	$p^* \times 100$	$\widehat{p} \times 100$	$F(T^*) \times 100$			
			$m = 1.6$	$m = 2.0$	$m = 2.4$	$m = 3.0$
2	38.2	33.3	91	70	55	41
4	27.6	20.0	46	30	22	16
6	22.2	14.3	30	19	14	10
10	16.5	9.1	18	11	8	5

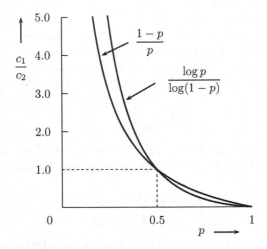

Fig. 9.11. Relationship with p and c_1/c_2

Thus, if we knew previously $\widehat{p}$ or p^* and c_2 by some method, then we could estimate the replacement cost c_1 roughly. Figure 9.11 presents the relationship with $\widehat{p}$, p^*, and c_1/c_2. The ratio of c_1/c_2 increases drastically as p tends to 0. It is clearly proved that $(1-p)/p > \log p/\log(1-p)$, i.e., $(1-p)\log(1-p) > p\log p$ for $0 < p < 1/2$.

Note that the relation

$$\frac{\log F(T)}{\log \overline{F}(T)} = \frac{c_1}{c_2}$$

is also derived from (9.20) by differentiating the function

$$\frac{-F(T)\log F(T) - \overline{F}(T)\log \overline{F}(T)}{c_1 F(T) + c_2 \overline{F}(T)}$$

with respect to T and setting it equal to zero.

(27) *Other maintenance models.* The entropy model can be applied to other maintenance models:

(i) $p = [F(T+T_0) - F(T_0)]/\overline{F}(T_0)$ for the age replacement in Sect. 5.2, where the unit is replaced at time $T_0 + T$ or at failure, whichever occurs first, given that it operates at time T_0.

(ii) $p = [F(t) - F(t - T)]/F(t)$ for the backward model in **(1)** of Sect. 5.5, where we go back to time T from t when the failure was detected at time t.

(iii) $p = \int_0^\infty \overline{W}(t)dF(t)^n$ for the scheduling problem of a parallel system with n units in Sect. 5.3. If p is determined, then an optimum number n^* can be estimated from p.

(iv) $p = [F(T_{k+1}) - F(T_k)]/\overline{F}(T_k)$ for the inspection model [1, p. 201], where the unit is checked at time T_k, given that it did not fail at time T_k. In this case, c_1 is the failure cost when the unit fails during $[T_k, T_{k+1}]$, and c_2 is one inspection cost.

(v) $p = G(T)$ for the repair limit model [1, p. 51], where a failed unit is repaired according to a repair distribution $G(t)$ and is replaced with a new one when its repair is not completed within time T. In this case, c_1 is the repair cost when the repair is completed until time T, and c_2 is the repair and replacement cost when the repair is not completed until time T.

10

Management Models with Reliability Applications

There exist many stochastic models in management science that have been studied by applying the techniques of reliability theory. Such models have appeared in the journals of Operations Research and Management Science and in books on probability and stochastic processes. This chapter surveys four recent studies of management models: (1) the definition of service reliability, (2) two optimization problems in the ATMs of a bank, (3) the loan interest rate of a bank, and (4) the CRL issue in PKI architecture.

Section 10.1 defines service reliability on hypothetical assumptions and investigates its properties [184]. It is shown that the service reliability function draws an upside-down bathtub curve. This would trigger beginning theoretical studies of service reliability in the near future. Section 10.2 takes up two optimization problems that are sometimes generated in ATMs of a bank [185, 186]: One is the maintenance of unmanned ATMs with two breakdowns, and the other is the number of spare cash-boxes for unmanned ATMs, where the cash-box is replaced with a new one when all the cash has been drawn out. The expected costs for two models are obtained, and optimum policies that minimize them are derived analytically.

Particularly in Japan, risk management relating to the bankruptcy of financed enterprises has become very important to a bank. Section 10.3 attempts to determine an adequate interest rate, taking account of the probabilities of bankruptcy and mortgage collection from practical viewpoints [187]. Finally, Section 10.4 presents optimum issue intervals for a certificate revocation list (CRL) in Public Key Infrastructure (PKI) [188,189]. Three models are proposed, and the expected costs for each model are obtained. These models are compared with each other.

Numerical examples are given in all sections to illustrate these models well and to understand their results easily. Furthermore, there exist a lot of similar models in the fields of management science and operations research. Such formulations and techniques shown in this chapter could be applied to actual models and be more useful for analyzing other similar stochastic models.

10.1 Service Reliability

The theory of software reliability [190, 191] has been highly developed apart from hardware reliability, as computers have spread widely to many fields and the demand for high reliability has increased greatly. From similar viewpoints, the theory of service reliability is beginning to be studied gradually: The case study of service dependability for transit systems was presented [192]. Some interesting methodologies of dealing with *service* from the point of engineering and of defining service reliability by investigating its qualities were proposed [193]. The methods for modeling service reliability of the logistic system in a supply chain were introduced [194]. Furthermore, different kinds of service in computer systems were considered: Software tools for evaluating serviceability [195], the service quality of failure detectors in distributed systems [196], and service reliability in grid systems [197, 198] were presented. Recently, International Service Availability Symposiums (ISAS) in dependable systems are held every year. There are many research and practical papers looking from different *services* related to areas in industry and academia.

A reliability function of service reliability has not yet been established theoretically. This section attempts to develop a theoretical approach to defining a new service reliability and to derive its reliability function based on our way of thinking.

(1) Service Reliability 1

It is assumed that service reliability is defined as the occurrence probabilities of the following two independent events:

Event 1: Service has N ($N = 0, 1, 2, \dots$) faults at the beginning that will occur successively and randomly, and its reliability improves gradually by removing them.

Event 2: Service goes down with time due to successive faults that occur randomly.

First, we derive the reliability function of Event 1: Suppose that N faults occur independently from time 0 according to an exponential distribution $(1 - e^{-\lambda_1 t})$ and are removed. We define the reliability function as $e^{-(N-k)\mu_1 t}$ when k ($k = 0, 1, 2, \dots, N$) faults have occurred in time t. Then, the reliability function for Event 1 is given by

$$R_1(t) = \sum_{k=0}^{N} e^{-(N-k)\mu_1 t} \binom{N}{k} \left(1 - e^{-\lambda_1 t}\right)^k \left(e^{-\lambda_1 t}\right)^{N-k}$$

$$= \left[1 - e^{-\lambda_1 t} + e^{-(\lambda_1 + \mu_1)t}\right]^N \quad (N = 0, 1, 2, \dots). \quad (10.1)$$

It is clearly seen that $R_1(0) = R_1(\infty) = 1$. Differentiating $R_1(t)$ with respect to t and setting it equal to zero,

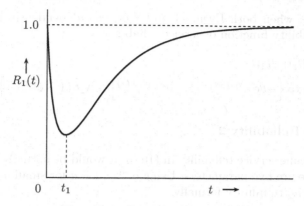

Fig. 10.1. General graph of $R_1(t)$

$$e^{-\mu_1 t} = \frac{\lambda_1}{\lambda_1 + \mu_1}. \tag{10.2}$$

Thus, $R_1(t)$ starts from 1 and has a minimum at $t_1 = (1/\mu_1) \log[(\lambda_1 + \mu_1)/\lambda_1]$, and after this, increases slowly to 1 (Fig. 10.1).

In general, service has a preparatory time to detect faults like a test time in software reliability. Thus, it is supposed that service starts initially after a preparatory time t_1. Setting t by $t + t_1$ in (10.1), the reliability function for Event 1 is

$$R_1(t) = \left[1 - e^{-\lambda_1(t+t_1)} + e^{-(\lambda_1+\mu_1)(t+t_1)}\right]^N. \tag{10.3}$$

If the number N of faults is a random variable according to a Poisson distribution with a mean θ, then (10.3) becomes

$$R_1(t) = \sum_{N=0}^{\infty} \left[1 - e^{-\lambda_1(t+t_1)} + e^{-(\lambda_1+\mu_1)(t+t_1)}\right]^N \frac{\theta^N}{N!} e^{-\theta}$$

$$= \exp\left\{-\theta e^{-\lambda_1(t+t_1)} \left[1 - e^{-\mu_1(t+t_1)}\right]\right\}, \tag{10.4}$$

that increases strictly from $R_1(0)$ to 1.

Next, suppose that faults of Event 2 occur according to a Poisson distribution with a mean λ_2, and its reliability function is defined by $e^{-k\mu_2 t}$ when k ($k = 0, 1, 2, \dots$) faults have occurred in time t. Then, the reliability function of Event 2 is given by

$$R_2(t) = \sum_{k=0}^{\infty} e^{-k\mu_2 t} \frac{(\lambda_2 t)^k}{k!} e^{-\lambda_2 t} = \exp\left[-\lambda_2 t \left(1 - e^{-\mu_2 t}\right)\right], \tag{10.5}$$

that decreases strictly from 1 to 0.

Therefore, when both Events 1 and 2 occur independently in series, we give the reliability function of service reliability as

$$R(t) \equiv R_1(t)R_2(t)$$
$$= \exp\left\{-\theta e^{-\lambda_1(t+t_1)}\left[1 - e^{-\mu_1(t+t_1)}\right] - \lambda_2 t\left(1 - e^{-\mu_2 t}\right)\right\}. \quad (10.6)$$

(2) Service Reliability 2

Even if we define service reliability in (10.6), it would be actually meaningless because there are five parameters. Using $e^{-a} \approx 1 - a$ for small a and tending both μ_1 and μ_2 to infinity formally,

$$R(t) \approx e^{-\lambda_2 t}\left[1 - \theta e^{-\lambda_1(t+t_1)}\right].$$

Thus, with reference to the above approximation, we define the reliability function of service reliability as

$$\widetilde{R}(t) \equiv \left(1 - \alpha e^{-\mu_1 t}\right)e^{-\mu_2 t} \quad (0 < \alpha < 1, \ 0 < \mu_1 < \infty, \ 0 < \mu_2 < \infty). \quad (10.7)$$

It is clearly seen that $\widetilde{R}(0) = 1 - \alpha$ and $\widetilde{R}(\infty) = 0$, $i.e.$, $1 - \alpha$ is estimated as an initial reliability. Furthermore, differentiating $\widetilde{R}(t)$ with respect to t and setting it equal to zero,

$$e^{-\mu_1 t} = \frac{1}{\alpha}\frac{\mu_2}{\mu_1 + \mu_2}.$$

Therefore, we have the following properties:

(i) If $\alpha > \mu_2/(\mu_1 + \mu_2)$, then $\widetilde{R}(t)$ starts from $1 - \alpha$ and has a maximum $\widetilde{R}(t_1) = [\mu_1/(\mu_1 + \mu_2)]e^{-\mu_2 t_1}$ at $t_1 = (-1/\mu_1)\log\{\mu_2/[\alpha(\mu_1 + \mu_2)]\}$, and after this, decreases to 0.

(ii) If $\alpha \leq \mu_2/(\mu_1 + \mu_2)$, then $\widetilde{R}(t)$ decreases strictly from $1 - \alpha$ to 0.

Figure 10.2 shows $\widetilde{R}(t)$ roughly for $\alpha > \mu_2/(\mu_1 + \mu_2)$.

A service reliability function would generally increase first and become constant for some interval, and after that, would decrease gradually, $i.e.$, it yields an upside-down bathtub curve [103]. The reliability function $\widetilde{R}(t)$ in Fig. 10.2 draws nearly such a curve. Furthermore, the preventive maintenance should be done at time t_2 that is given by the solution of $\widetilde{R}(t) = 1 - \alpha$, $i.e.$,

$$\frac{1 - e^{-\mu_2 t}}{1 - e^{-(\mu_1 + \mu_2)t}} = \alpha.$$

A service reliability function could be investigated from many angles of service approaches and be verified to have a general curve such as the bathtub in reliability theory.

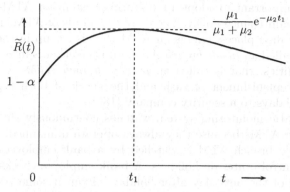

Fig. 10.2. General graph of $\widetilde{R}(t)$

10.2 Optimization Problems in ATMs

There exist recently many unmanned automatic teller machines (ATMs) in banks that customers can use evens on holidays. An automatic monitoring system continuously watches the operation of ATMs through telecommunication network to prevent some problems. There are mainly two kinds problems found in ATMs: One occurs inside the branch, where ATMs are manned except on weekends and holidays, and the other occurs outside the branch, where ATMs always operate unmanned. Two kinds of breakdowns are considered, and the expected cost for an unmanned operating period is obtained. A maintenance policy that minimizes the expected cost is derived analytically.

When all the cash in an ATM has been drawn out, the cash-box is replaced with one of spares. Next, we consider the problem of how many cash-boxes should be provided at the beginning. The total expected cost is derived by introducing several costs incurred for the ATM operation, and an optimum number that minimizes it is discussed.

10.2.1 Maintenance of ATMs

Automatic tellers machines (ATMs) in banks have rapidly spread to our daily life, and also, their operating hours have increased greatly. Recently, some ATMs are needed to be open even on weekends and holidays according to customers' demand. ATMs have various kinds of facilities such as the transfer of cash, the contact and cancellation of deposits and accounts, the loan payments, and so on. Most ATMs are connected with the online system of a bank and increase the efficiency of business. Furthermore, ATMs are also connected with other organizations whose networks would be expanded in every nook and corner. In such situations, adequate and prompt maintenance for some problems and breakdowns of ATMs has to be done from both viewpoints of customers' trust and service.

It is very important to adopt a monitoring system for ATMs and to plan previously a maintenance policy. There are two kinds of ATMs in accordance with their installed places. One is an ATM that is set up in the branch of a bank, that is called an *inside branch ATM*, and the other is in stores, stations, or public facilities, that is called an *outside branch ATM*. A bank usually consigns the replenishment of cash and the check of both types of ATMs except on weekdays to a security company [186].

An automatic monitoring system watches continuously the operation of outside branch ATMs because they always operate unmanned. On the other hand, an inside branch ATM is watched by a bank employee on weekdays and by the control center on holidays. A bank employee checks the ATM at the beginning of the next day after holidays. Even if some problems occur in the ATM on holidays, they are removed by a bank employee on the next day, and the ATM is restored to a normal condition. A monitoring system at the control center can display problems for outside branch ATMs in the terminal unit and output them. Moreover, there might sometimes be phone calls from users in ATMs to report problems. If such problems are displayed at the terminal unit, a worker at the control center can remove some of them by operating the terminal unit remotely according to their states. Otherwise, a worker reports such facts to a security company that can correct promptly ATM problems or breakdowns.

Suppose that there exist two kinds of breakdowns by which ATMs break down after trouble occurs. We propose a stochastic model for an inside branch ATM with two breakdowns that operates unmanned on a weekend and is checked after trouble occurs. This is one kind of modified inspection models [1, p. 201], where an operating unit is checked at planned times to inspect whether or not it has failed. The probability distributions of times to each occurrence of two breakdowns are given, and the checking and breakdown costs are introduced. Then, the expected cost for ATMs for an unmanned operating period is obtained, and an optimum maintenance policy that minimizes it is derived analytically.

(1) Expected Cost

An automatic monitoring system watches an inside branch ATM on holidays through a telephone line and can display its state. The state is generally classified into the following five groups:

State 0 : The ATM is normal.

State 1 : There are some troubles in the ATM such that the cash and receipts may be running out soon, or the ATM may be choked up with cards and cash. The ATM will break down soon because of these troubles. If a worker at the control center can remove such troubles remotely, they are not included in State 1.

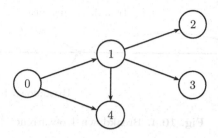

Fig. 10.3. Figure of transient among five states

State 2: The ATM is checked at time T after State 1. A security company worker runs to the ATM and can remove troubles before its breakdown. This is an easy job that requires changing the cash-box or replenishing the receipts and journal forms.

State 3: The ATM breaks down until time T after State 1 (Breakdown 1), *i.e.*, it breaks down before a security worker arrives at the ATM. The worker recovers the ATM by changing the cash-box or replenishing the receipts and journal forms.

State 4: The ATM breaks down due to mechanical causes such as the power supply stops and is choked up with cash and cards, and so on. A security worker runs to the ATM and recovers it. The ATM cannot be used from the breakdown to the arrival time of the worker. The maintenance time for Breakdown 2 would be longer than that for Breakdown 1 in State 3.

Figure 10.3 shows the transition relation among the five states.

In the operation of an ATM, some troubles with cash, receipt forms, and journal forms would occur at most once for a short time such as a weekend or a holiday. Suppose that the ATM has to operate during the interval $[0, S]$, and the trouble occurs only at most once in $[0, S]$. It is assumed that some trouble occurs according to a general distribution $F_0(t)$, and after its occurrence, the time to Breakdown i $(i = 1, 2)$ has a general distribution $F_i(t)$. The trouble and two breakdowns occur independently of each other. If there are two or more ATMs in the same booth, five states are denoted as the state of the last operating ATM.

We get the following probabilities that events such as troubles and breakdowns occur in $[0, S]$, where $\overline{F}_i(t) \equiv 1 - F_i(t)$:

(1) The probability that any troubles and Breakdown 2 do not occur in $[0, S]$ is

$$\overline{F}_0(S)\overline{F}_2(S). \tag{10.8}$$

(2) The probability that Breakdown 2 occurs before trouble in $[0, S]$ is

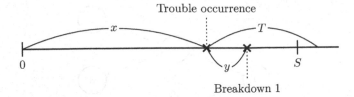

Fig. 10.4. Breakdown 1 occurrence

$$\int_0^S \overline{F}_0(x)\, \mathrm{d}F_2(x). \tag{10.9}$$

(3) The probability that the ATM is checked at time S without breakdowns after trouble is

$$\overline{F}_2(S) \int_{S-T}^S \overline{F}_1(S - x)\, \mathrm{d}F_0(x). \tag{10.10}$$

(4) The probability that Breakdown 1 occurs after trouble (Fig. 10.4) is

$$\int_{S-T}^S \left[\int_0^{S-x} \overline{F}_2(x + y)\mathrm{d}F_1(y)\right] \mathrm{d}F_0(x). \tag{10.11}$$

(5) The probability that Breakdown 2 occurs after trouble is

$$\int_{S-T}^S \left[\int_x^S \overline{F}_1(y - x)\mathrm{d}F_2(y)\right] \mathrm{d}F_0(x). \tag{10.12}$$

(6) The probability that the ATM is checked at time T after trouble is

$$\overline{F}_1(T) \int_0^{S-T} \overline{F}_2(T + x)\, \mathrm{d}F_0(x). \tag{10.13}$$

(7) The probability that Breakdown 1 occurs until time T after trouble is

$$\int_0^{S-T} \left[\int_0^T \overline{F}_2(x + y)\mathrm{d}F_1(y)\right] \mathrm{d}F_0(x). \tag{10.14}$$

(8) The probability that Breakdown 2 occurs until time T after trouble (Fig. 10.5) is

$$\int_0^{S-T} \left[\int_x^{x+T} \overline{F}_1(y - x)\mathrm{d}F_2(y)\right] \mathrm{d}F_0(x). \tag{10.15}$$

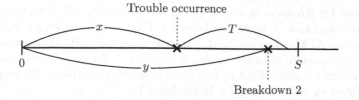

Fig. 10.5. Breakdown 2 occurrence

Clearly

$$(10.10) + (10.11) + (10.12)$$

$$= \int_{S-T}^{S} \left[\overline{F}_2(S)\overline{F}_1(S-x) + \int_0^{S-x} \overline{F}_2(x+y)\,\mathrm{d}F_1(y) \right.$$

$$\left. + \int_x^S \overline{F}_1(y-x)\,\mathrm{d}F_2(y) \right] \mathrm{d}F_0(x)$$

$$= \int_{S-T}^{S} \overline{F}_2(x)\,\mathrm{d}F_0(x). \tag{10.16}$$

$$(10.13) + (10.14) + (10.15)$$

$$= \int_0^{S-T} \left[\overline{F}_2(T+x)\overline{F}_1(T) + \int_0^T \overline{F}_2(x+y)\,\mathrm{d}F_1(y) \right.$$

$$\left. + \int_x^{x+T} \overline{F}_1(y-x)\mathrm{d}F_2(y) \right] \mathrm{d}F_0(x)$$

$$= \int_0^{S-T} \overline{F}_2(x)\,\mathrm{d}F_0(x). \tag{10.17}$$

Thus, it follows that

$$(10.8) + (10.9) + (10.16) + (10.17)$$

$$= \overline{F}_0(S)\overline{F}_2(S) + \int_0^S \overline{F}_0(x)\,\mathrm{d}F_2(x) + \int_0^S \overline{F}_2(x)\,\mathrm{d}F_0(x) = 1.$$

We introduce the following costs:

$c_0 =$ Cost when the ATM stops at time S. A bank employee checks the ATM before it begins to operate on the next day and replenishes the cash, journal, and receipt forms.

$c_1 =$ Checking cost at time T after trouble. A security worker refills the cash-box, and if necessary, replenishes the journal and receipt forms. Cost c_1 is higher than c_0 because a security worker has to go to the ATM.

$c_2 =$ Cost for Breakdown 1. The ATM has stopped until a security worker arrives at time T after Breakdown 1. Customers cannot use it and have to use ATMs of other banks. In this case, not only do customers pay a commission to other banks, but also a bank pays a commission for customers' usage. Cost c_2 includes the whole cost that is the summation of cost c_1 and the cost for Breakdown 1.

$c_3 =$ Cost for Breakdown 2. The ATM breaks down directly and has stopped until a security worker arrives at the ATM. The maintenance time and cost for Breakdown 2 would be usually be longer and higher than those for Breakdown 1, respectively. It can be seen in general that $c_3 > c_2 > c_1 > c_0$.

From the notations of the above costs, the total expected cost for ATM operation during the interval $[0, S]$ is

$$
\begin{aligned}
C(T) =\ & c_0 \times [(10.8) + (10.10)] + c_1 \times (10.13) + c_2 \times [(10.11) + (10.14)] \\
& + c_3 \times [(10.9) + (10.12) + (10.15)] \\
=\ & c_0 \overline{F}_2(S) \left[\overline{F}_0(S) + \int_{S-T}^{S} \overline{F}_1(S - x)\, \mathrm{d}F_0(x) \right] \\
& + c_1 \overline{F}_1(T) \int_0^{S-T} \overline{F}_2(T + x)\, \mathrm{d}F_0(x) \\
& + c_2 \left\{ \int_{S-T}^{S} \left[\int_0^{S-x} \overline{F}_2(x + y)\, \mathrm{d}F_1(y) \right] \mathrm{d}F_0(x) \right. \\
& \left. + \int_0^{S-T} \left[\int_0^{T} \overline{F}_2(x + y)\, \mathrm{d}F_1(y) \right] \mathrm{d}F_0(x) \right\} \\
& + c_3 \left\{ \int_0^{S} \overline{F}_0(x)\, \mathrm{d}F_2(x) + \int_{S-T}^{S} \left[\int_x^{S} \overline{F}_1(y - x)\, \mathrm{d}F_2(y) \right] \mathrm{d}F_0(x) \right. \\
& \left. + \int_0^{S-T} \left[\int_x^{x+T} \overline{F}_1(y - x)\, \mathrm{d}F_2(y) \right] \mathrm{d}F_0(x) \right\}.
\end{aligned}
\tag{10.18}
$$

(2) Optimum Policy

It is a problem to determine when a security worker should go to the ATM after trouble occurs. For example, if trouble occurs in near time S, it would be unnecessary to send a security worker. We find an optimum time T^* $(0 \leq T^* \leq S)$ that minimizes the expected cost $C(T)$ in (10.18). In the particular case of $T = 0$, $i.e.$, when the ATM is maintained immediately after trouble, the expected cost is

$$
C(0) = c_0 \overline{F}_2(S) \overline{F}_0(S) + c_1 \int_0^{S} \overline{F}_2(x)\, \mathrm{d}F_0(x) + c_3 \int_0^{S} \overline{F}_0(x)\, \mathrm{d}F_2(x).
\tag{10.19}
$$

In the particular case of $T = S$, $i.e.$, when the ATM is not maintained until time S even if troubles occur, the expected cost is

$$C(S) = c_0 \overline{F}_2(S) \left[\overline{F}_0(S) + \int_0^S \overline{F}_1(S-x) \, dF_0(x) \right]$$

$$+ c_2 \int_0^S \left[\int_0^{S-x} \overline{F}_2(x+y) \, dF_1(y) \right] dF_0(x)$$

$$+ c_3 \left\{ \int_0^S \overline{F}_0(x) \, dF_2(x) + \int_0^S \left[\int_x^S \overline{F}_1(y-x) \, dF_2(y) \right] dF_0(x) \right\}.$$

$$(10.20)$$

Next, suppose that $F_0(t)$ and $F_2(t)$ are exponential, $i.e.$, $F_0(t) = 1 - e^{-\lambda_0 t}$ and $F_2(t) = 1 - e^{-\lambda_2 t}$. In addition, assume that $F_1(t)$ has a density function $f_1(t)$, and define its failure rate as $h_1(t) \equiv f_1(t)/\overline{F}_1(t)$. Differentiating $C(T)$ in (10.18) with respect to T and setting it equal to zero,

$$[(c_2 - c_1)h_1(T) + (c_3 - c_1)\lambda_2] \frac{e^{(\lambda_0 + \lambda_2)(S-T)} - 1}{\lambda_0 + \lambda_2} = c_1 - c_0. \qquad (10.21)$$

In general, it would be very difficult to derive an optimum time T^* analytically. We give the following results that are useful for computing T^* numerically.

(i) If there exists a solution to satisfy (10.21), then an optimum time is given by comparing $C(0)$ in (10.19), $C(S)$ in (10.20), and $C(T)$ in (10.18).
(ii) If there is no solution to satisfy (10.21), an optimum time is $T^* = S$ because $C(T)$ decreases with T.

Recently, ATMs have developed rapidly and are set up in many different places. It would be necessary to plan appropriate maintenance for each of ATMs by modifying this model and estimating statistically suitable parameters from actual ones.

10.2.2 Number of Spare Cash-boxes

Most banks in the branch have set up many unmanned ATMs, where customers can deposit or withdraw money even on a weekend and holidays. There are some small cash-boxes that hold cash in each ATM.

There are usually two different types of boxes in ATMs. One is a box only for deposits and the other is a box only for withdrawing. According to customers' demand, a constant amount of cash is kept beforehand in each cash-box. If all the cash in a box has been drawn out, the ATM operation is stopped. Then, the ATM service is restarted by replacing an empty box with a new one.

When a cash-box becomes empty on weekdays, a banker can usually replenish it with cash. However, when all the cash in an ATM has been drawn out on holidays, a security company receives the information from the control center that continuously monitors ATMs and replaces it quickly with a new one. Thereafter, the service for drawing cash begins again. This company provides in advance some spare cash-boxes for such situations. Such replacements may be repeated during a day.

One important problem arising in the above situation is how many spare cash-boxes per each branch should be provided to a security company. As one method of answering this question, we introduce some costs and form a stochastic model. We derive the expected cost and determine an optimum number of spare cash-boxes that minimizes it. This is one modification of discrete replacement models [1, 35]. The methods used in this section would be applied to the maintenance of other automatic vending machines.

(1) Expected Cost

We examine one ATM in the branch of a bank. When there are several ATMs in the branch and all the cash of the boxes has been withdrawn from all ATMs, the company replaces them with only one new box. Thus, we use the word of ATM without ATMs. The costs might be mainly incurred in the following three cases for the ATM operation:

(1) The cash in spare cash-boxes is surplus funds and brings no profit if it is not used.
(2) When all the cash in the ATM has been drawn out, customers would draw cash from other ATMs of the bank or other banks. In this case, if customers use ATMs of other banks, not only would they have to pay the extra commission, but also a bank has to allow some commissions to other banks. Conversely, if customers of other banks use this ATM, the bank can receive a commission from customers and other banks.
(3) The bank has to pay a fixed contract deposit, and also, pay a constant commission to a security company, whenever the company delivers a spare cash-box and exchanges it for an empty one.

From the above viewpoints (1), (2), and (3), we introduce the following three costs: All the cash provided in spare cash-boxes incurs an opportunity cost c_1 per unit of cash, and when customers use other banks, this incurs cost c_2 per unit of cash. In addition, cost c_3 is needed for each exchange of one box, whenever a security company delivers a spare cash-box.

Suppose that $F(x)$ is the distribution function of the total amount of cash that is drawn each day from the ATM in the branch, and its mean is $\mu \equiv \int_0^\infty \overline{F}(x)\mathrm{d}x < \infty$, where $\overline{F}(x) \equiv 1 - F(x)$. Let α $(0 \le \alpha < 1)$ be the rate at which customers of other banks have drawn cash from the ATM and β $(0 \le \beta < 1)$ be the probability that customers give up drawing cash or use other ATMs of the bank, *i.e.*, $1 - \beta$ is the probability that they use ATMs of other banks when all ATMs in the branch have stopped.

It is assumed that N is the number of spare cash-boxes in the ATM, and a is the amount of cash stored in one cash-box. Thus, the first amount of cash stored in the ATM is $A \equiv Na$, and hence, the total cost required for providing N spare cash-boxes is

$$c_1 Na. \tag{10.22}$$

The cost for the case where customers use ATMs of other banks, when the ATM has stopped, is

$$c_2(1 - \alpha)(1 - \beta) \int_{A+Na}^{\infty} (x - A - Na) \, dF(x). \tag{10.23}$$

Conversely, the profit paid for the bank by customers of other banks who use the ATM is

$$-c_2\alpha \left[(A + Na)\overline{F}(A + Na) + \int_0^{A+Na} x \, dF(x) \right] = -c_2\alpha \int_0^{A+Na} \overline{F}(x) \, dx. \tag{10.24}$$

This also includes the commission that customers pay to the branch. The above formulations of $(10.22) - (10.24)$ are cost functions similar to the classical *Newspaper sellers problem* [199]. In addition, the total cost required for a security company that delivers spare cash-boxes is

$$c_3 \left[\sum_{j=0}^{N-1} (j + 1) \int_{A+ja}^{A+(j+1)a} dF(x) + N \int_{A+Na}^{\infty} dF(x) \right] = c_3 \sum_{j=0}^{N-1} \overline{F}(A + ja), \tag{10.25}$$

where $\sum_{j=0}^{-1} \equiv 0$. Summing up $(10.22) - (10.25)$ and arranging it, the total expected cost $C(N)$ is given by

$$C(N) = c_1 Na + c_2 \left\{ [1 - (1 - \alpha)\beta] \int_{A+Na}^{\infty} \overline{F}(x) \, dx - \alpha\mu \right\} + c_3 \sum_{j=0}^{N-1} \overline{F}(A + ja)$$

$$(N = 0, 1, 2, \dots). \tag{10.26}$$

(2) Optimum Policy

We find an optimum number N^* of spare cash-boxes that minimizes $C(N)$ in (10.26). It is clearly seen that $C(\infty) \equiv \lim_{N \to \infty} C(N) = \infty$,

$$C(0) = c_2 \left\{ [1 - (1 - \alpha)\beta] \int_A^{\infty} \overline{F}(x) \, dx - \alpha\mu \right\}. \tag{10.27}$$

Thus, there exists a finite N^* $(0 \leq N^* < \infty)$ that minimizes $C(N)$. Forming the inequality $C(N + 1) - C(N) \geq 0$ to seek an optimum number N^*,

$$\frac{c_1 a + c_3 \overline{F}(A + Na)}{\int_{A+Na}^{A+(N+1)a} \overline{F}(x)\mathrm{d}x} \geq c_2 \gamma, \qquad (10.28)$$

where $\gamma \equiv 1 - (1 - \alpha)\beta > 0$.

Assume that a density function $f(x)$ of $F(x)$ exists. Let us denote the left-hand side of (10.28) by $L(y)$, where $y \equiv A + Na$, and investigate the properties of $L(y)$ that is given by

$$L(y) \equiv \frac{c_1 a + c_3 \overline{F}(y)}{\int_y^{y+a} \overline{F}(x)\mathrm{d}x} \qquad (y \geq A). \qquad (10.29)$$

It is clear that $L(\infty) \equiv \lim_{y \to \infty} L(y) = \infty$,

$$L(A) = \frac{c_1 a + c_3 \overline{F}(A)}{\int_A^{A+a} \overline{F}(x)\mathrm{d}x},$$

$$\frac{\mathrm{d}L(y)}{\mathrm{d}y} = \frac{F(y + a) - F(y)}{\int_y^{y+a} \overline{F}(x)\mathrm{d}x} \left[\frac{c_1 a + c_3 \overline{F}(y)}{\int_y^{y+a} \overline{F}(x)\mathrm{d}x} - \frac{c_3 f(y)}{F(y + a) - F(y)} \right].$$

If $F(x)$ has the property of IFR (Increasing Failure Rate), i.e., the failure rate $f(x)/\overline{F}(x)$ increases, then

$$\frac{f(x)}{\overline{F}(x)} \geq \frac{f(y)}{\overline{F}(y)} \qquad (y \leq x \leq y + a).$$

Hence,

$$L(y) \geq \frac{c_1 a + c_3 \overline{F}(y)}{\int_y^{y+a} f(x)[\overline{F}(y)/f(y)]\mathrm{d}x} = \frac{\{[c_1 a / \overline{F}(y)] + c_3\} f(y)}{F(y + a) - F(y)}$$

$$> \frac{c_3 f(y)}{F(y + a) - F(y)}.$$

Thus, if $F(x)$ is IFR, then $L(y)$ increases strictly from $L(A)$ to ∞.

Therefore, we can give the following optimum number N^* when $F(x)$ is IFR:

(i) If $L(A) < c_2 \gamma$, then there exists a finite and unique minimum $N^*(1 \leq N^* < \infty)$ that satisfies (10.28).

(ii) If $L(A) \geq c_2 \gamma$, then $N^* = 0$, i.e., we should provide no spare cash-box.

We can explain the reason why $F(x)$ has the property of IFR: The total amount of cash on ordinary days is almost constant, however, a lot of money is drawn out at the end of the month just after most workers have received their salaries. It is well-known that its amount is about 1.75 times more than that of ordinary days. Thus, we might consider that the drawing rate of cash is

Table 10.1. Optimum number N^* and expected cost $C(N^*)$ when $A = a = 20.0$, $\alpha = 0.4$, $\beta = 0.6$, $1/\lambda = 25.0$, and $c_3 = 0.01$

c_2	c_1			
	0.000068		0.000164	
	N^*	$C(N^*)$	N^*	$C(N^*)$
0.001	0	-0.00325	0	-0.00325
0.002	1	-0.00753	0	-0.00650
0.003	2	-0.01456	1	-0.01194
0.004	2	-0.02205	1	-0.01828
0.005	3	-0.02983	2	-0.02572

constant or increases with the total amount of cash. Note that both exponential and normal distributions given in numerical examples have the property of IFR.

Example 10.1. Consider two cases where $F(x)$ has exponential and normal distributions. First, suppose that $F(x) = 1 - e^{-\lambda x}$ for $x \geq 0$. Then, (10.26) and (10.28) are rewritten as, respectively,

$$C(N) = c_1 N a + \frac{c_2}{\lambda}\left[\gamma e^{-\lambda(A+Na)} - \alpha\right] + c_3 e^{-\lambda A}\frac{1 - e^{-N\lambda a}}{1 - e^{-\lambda a}}, \qquad (10.30)$$

$$c_1 a e^{\lambda(A+Na)} \geq c_2 \gamma\frac{1 - e^{-\lambda a}}{\lambda} - c_3. \qquad (10.31)$$

Therefore, the optimum policy is as follows:

(i) If $c_1 a e^{\lambda A} + c_3 < c_2 \gamma\left[(1 - e^{-\lambda a})/\lambda\right]$, then there exists a unique minimum $N^*(1 \leq N^* < \infty)$ that satisfies (10.31).
(ii) If $c_1 a e^{\lambda A} + c_3 \geq c_2 \gamma\left[(1 - e^{-\lambda a})/\lambda\right]$, then $N^* = 0$.

In the case of (i), by solving (10.31) with respect to N,

$$N^* = \left[\frac{1}{\lambda a}\log\left(\frac{c_2\gamma}{c_1 a}\frac{1 - e^{-\lambda a}}{\lambda} - \frac{c_3}{c_1 a}\right) - \frac{A}{a}\right] + 1, \qquad (10.32)$$

where $[x]$ denotes the greatest integer contained in x. It is clearly seen from (10.31) or (10.32) that an optimum N^* increases with c_2/c_1 and decreases with c_3/c_1. In the case of (ii), note that if

$$c_1 a(1 + \lambda A) + c_3 \geq c_2 a \gamma, \qquad (10.33)$$

then $N^* = 0$.

Suppose that $A = a = 20.0$, where a denominator of money is 1 million yen $\risingdotseq \$9,000$, and the yields on investment are $c_1 = 2.5$ or 6.0% per year, *i.e.*,

Table 10.2. Upper number $\overline{N}$ when $A = a = 20.0$

ε	$1/\lambda$	
	20.0	30.0
0.20	1	2
0.10	2	3
0.05	2	4

$c_1 = 0.025/365 = 0.000068$ or $0.06/365 = 0.000164$ per day, and $c_3 = 10,000$ yen $=0.01$. In addition, the mean of the total cash each day is $1/\lambda = 25.0$. Then, from (10.32) and (10.33), we can compute the optimum number N^* and the resulting cost $C(N^*)$ for a given c_2. Table 10.1 presents the computing results N^* and $C(N^*)$ for $c_2 = 1,000$–$5,000$ yen per $100,000$ yen, $i.e.$, $c_2 = 0.001$–0.005. For example, when $c_1 = 2.5\%$ per year, $c_2 = 3,000$ yen, and $c_3 = 10,000$ yen, $N^* = 2$ and $C(N^*) = -14,560$ yen. In this case, we should provide two spare cash-boxes, and a bank gains a profit of 14,560 yen per day. In addition, the probability that the total cash has been drawn out is

$$e^{-(20.0+2\times 20.0)/25.0} \fallingdotseq 0.0907.$$

It is clearly seen that N^* increases with c_2, and $C(N^*)$ is negative for $N^* > 0$ and decreases with c_2 because it yields a profit when customers of other banks use the ATM. If c_1 is large, $C(N^*)$ increases, $i.e.$, the profit of bank decreases.

Moreover, we are interested in an upper spare number $\overline{N}$ in which the probability that the total cash has been drawn out a day from ATM is less than or equal to a small $\varepsilon > 0$. The upper number $\overline{N}$ is given by

$$\overline{F}(A + Na) = e^{-\lambda(A+Na)} \le \varepsilon. \tag{10.34}$$

Table 10.2 presents the upper number $\overline{N}$ for $1/\lambda = 20.0$, 30.0 and $\varepsilon = 0.20$, 0.10, and 0.05 when $A = a = 20.0$. It can be shown from this table how many spare cash boxes should be provided for a given ε.

Next, suppose that $\overline{F}(x) = [1/(\sqrt{2\pi}\sigma)]\int_x^\infty \exp[-(t-\mu)^2/(2\sigma^2)]\,dt$. Then, from (10.28), an optimum number N^* is given by a minimum number that satisfies

$$\frac{c_1 a + c_3\left[1/(\sqrt{2\pi}\sigma)\right]\int_{A+Na}^\infty \exp\left[-(t-\mu)^2/(2\sigma^2)\right]dt}{[1/(\sqrt{2\pi}\sigma)]\int_{A+Na}^{A+(N+1)a}\left\{\int_x^\infty \exp\left[-(t-\mu)^2/(2\sigma^2)\right]dt\right\}dx} \ge c_2\gamma. \tag{10.35}$$

Table 10.3 presents the optimum number N^* and the resulting cost $C(N^*)$ for $c_1 = 0.000068, 0.000164$ and $c_2 = 0.001$–0.005 when $A = a = 20.0$, $\alpha = 0.4$, $\beta = 0.6$, $\mu = 25.0$, $\sigma = 20.0$, and $c_3 = 0.01$. This indicates that the values of N^* are greater than those in Table 10.1 for $c_2 \ge 0.002$, and the resulting costs

Table 10.3. Optimum number N^* and expected cost $C(N^*)$ when $A = a = 20.0$, $\alpha = 0.4$, $\beta = 0.6$, $\mu = 25.0$, $\sigma = 20.0$, and $c_3 = 0.01$

c_2	c_1			
	0.000068		0.000164	
	N^*	$C(N^*)$	N^*	$C(N^*)$
0.001	0	0.16163	0	0.16163
0.002	3	-0.00666	3	-0.00090
0.003	3	-0.01637	3	-0.01061
0.004	3	-0.02608	3	-0.02033
0.005	4	-0.03580	3	-0.03004

Table 10.4. Upper number $\overline{N}$ when $A = a = 20.0$, $\sigma = 20.0$

ε	$\mu = 20.0$		$\mu = 30.0$	
	$\sigma = 20.0$	$\sigma = 30.0$	$\sigma = 20.0$	$\sigma = 30.0$
0.20	1	2	2	2
0.10	2	2	2	3
0.05	2	3	3	3

$C(N^*)$ are greater for small c_2 and are less for large c_2 than those in Table 10.1. However, the two tables have similar tendencies. It should be estimated from actual data which distribution is suitable for distribution $F(x)$.

We can compute an upper spare number $\overline{N}$ by the method similar to Table 10.2. This upper number $\overline{N}$ is given by

$$\overline{F}(A + Na) = \frac{1}{\sqrt{2\pi}\sigma} \int_{A+Na}^{\infty} \exp\left[-(t - \mu)^2/(2\sigma^2)\right] dt \le \varepsilon. \qquad (10.36)$$

Table 10.4 presents the upper number $\overline{N}$ for $\mu = 20.0, 30.0$, $\sigma = 20.0, 30.0$ and $\varepsilon = 0.20, 0.10$, and 0.05 when $A = a = 20.0$. It is natural that $\overline{N}$ increases with μ and σ. These values are equal to or less than those in Table 10.2 for $\mu = \sigma = 1/\lambda$. ∎

10.3 Loan Interest Rate

It would be necessary to consider the risk management of a bank from the viewpoint of credit risk. A bank has to lend at a high interest rate to enterprises with high risk from the well-known law of *high risk and high return* as market mechanism. When financed enterprises have become bankrupt, a bank has to collect the amount of loans from them as much as it can and to gain the earnings that correspond to the risk. However, the mortgage collection cost might sometimes be higher than the clerical working cost at the inception of loans. Moreover, the mortgage might be collected at once or at many times.

This section considers the following stochastic model of mortgage collection: The total amount of loans and mortgages can be collected in a batch at one time. A financed enterprise becomes bankrupt according to a bankruptcy probability and its mortgage collection probability. It is assumed that these probabilities are previously estimated and already known.

There have been many research papers that treat the determination of a loan interest rate considering default-risk and asset portfolios [200–208]. However, there have been few papers that study theoretically the period of mortgage collection. For example, the optimum duration of the collection of defaulted loans that maximizes the expected net profit was determined [209]. In this section, we are concerned with both mortgage collection time and loan interest rate when enterprises have become bankrupt.

A bank considers the loss of bankruptcy and the cost of mortgage collection and should decide a loan interest rate to gain earnings. In such situations, we formulate a stochastic model by using reliability theory, and discuss theoretically and numerically how to decide on an adequate loan interest rate.

10.3.1 Loan Model

In the interest rate model, the frequency of the compound interest is assumed to be infinity to use the differentiation. Continuous compound interest is called a *momentary interest rate*.

Suppose that P_0 is the present value of principal at time 0, and $P(t)$ is the total value of principal at time t including the interest. Then, if the interest increases with a momentary interest rate α, then the total interest is

$$P(t + \Delta t) = P(t)(1 + \alpha \Delta t) + o(\Delta t),$$

where $\lim_{\Delta t \to 0} o(\Delta t)/\Delta t = 0$. Thus, we have the following differential equation:

$$\frac{dP(t)}{dt} = \alpha P(t). \tag{10.37}$$

Solving this equation under the initial condition $P(0) = P_0$,

$$P_0 = P(t)e^{-\alpha t}. \tag{10.38}$$

When the momentary repayment μ per unit of time is consecutive for period t, P_0 is given by

$$P_0 = \int_0^t \mu e^{-\alpha u}\, du = \mu \frac{1 - e^{-\alpha t}}{\alpha}. \tag{10.39}$$

Thus,

$$\mu = P_0 \frac{\alpha}{1 - e^{-\alpha t}}. \tag{10.40}$$

Suppose that the bankruptcy probability of the financed enterprise is already known. We investigate the relation between the financing period and the interest rate for the installment repayment of the loan. The following notations are used:

M : All amounts of loans that can be procured by the deposit.

α_1 : Loan interest rate.

α_2 : Deposit interest rate with $\alpha_2 < \alpha_1$.

β : Ratio of mortgage to the amount of loans, i.e., ratio of the expected mortgage collection to the amount of loans ($0 < \beta \leq 1$).

$F(t)$, $f(t)$: Bankruptcy probability distribution and its density function, i.e., $F(t) \equiv \int_0^t f(u)du$, where $\overline{F}(t) \equiv 1 - F(t)$.

$Z(t)$, $z(t)$: Mortgage collection probability distribution and its density function, i.e., $Z(t) \equiv \int_0^t z(u)du$, where $\overline{Z}(t) \equiv 1 - Z(t)$.

It is assumed that all amounts of principal of the mortgage can be collected in a batch at one time. In this case, the distribution $Z(t)$ represents the probability that the principal of the mortgage can be collected until time t after the enterprise has become bankrupt.

(1) Expected Earning in Case of No Bankruptcy

In the installment repayment type of loans, let α_1 be a momentary interest rate, T be the financed period when no financed emprise becomes bankrupt, and M be the amount of loans for the financed enterprise at time 0. Then, the momentary repayment μ per unit of time is, from (10.40),

$$\mu = M \frac{\alpha_1}{1 - e^{-\alpha_1 T}}.$$

Thus, the total amount $S_1(T)$ of principal and loan interest is

$$S_1(T) = \int_0^T \mu e^{\alpha_1 (T-u)} \, du = M e^{\alpha_1 T}. \tag{10.41}$$

On the other hand, let α_2 be a momentary deposit interest rate, M be the amount of the deposit, and θ be the amount of withdraw per unit of time. Then, from (10.40),

$$\theta = M \frac{\alpha_2}{1 - e^{-\alpha_2 T}}.$$

Thus, the total amount $S_2(T)$ of the deposit at time T is

$$S_2(T) = \int_0^T \theta e^{\alpha_2 (T-u)} \, du = M e^{\alpha_2 T}. \tag{10.42}$$

Therefore, the earning of a bank at time T is, from (10.41) and (10.42),

$$P_1(T) \equiv S_1(T) - S_2(T) = M \left(e^{\alpha_1 T} - e^{\alpha_2 T} \right). \tag{10.43}$$

(2) Mortgage Collection Time

When a financed enterprise has become bankrupt at time t_0, the bank cannot collect the amount $(1 - \beta)Me^{\alpha_1 t_0}$, however, it can do $\beta Me^{\alpha_1 t_0}$ according to the mortgage collection probability distribution $Z(t - t_0)$ for $t \geq t_0$. But, it might be unprofitable to continue such collection until its completion.

Suppose that the mortgage collection stops at time t ($t \geq t_0$). We may consider that the clerical work for the mortgage collection would be almost the same whether the amount of mortgage is small or large, that is, it would be reasonable to assume that the clerical cost for the mortgage collection is constant regardless of the amount of the mortgage and is proportional to the working time. Thus, let c_1 be the constant cost that is proportional to both the amount of loans and the time of mortgage collection. Then, the expected earning $Q(t|t_0)$ from mortgage collection, when the enterprise became bankrupt at time t_0, is given by

$$
Q(t \mid t_0) = \int_{t_0}^{t} \left[\beta M e^{\alpha_1 t_0} - c_1(u - t_0) \right] dZ(u - t_0) - c_1(t - t_0)\overline{Z}(t - t_0)
$$

$$
= \beta M e^{\alpha_1 t_0} Z(t - t_0) - c_1 \int_0^{t-t_0} \overline{Z}(u) \, du. \tag{10.44}
$$

We find an optimum stopping time t^* ($t^* \geq t_0$) for mortgage collection that maximizes $Q(t \mid t_0)$ for a given t_0. Differentiating $Q(t \mid t_0)$ with respect to t and setting it equal to zero,

$$
\frac{z(t - t_0)}{\overline{Z}(t - t_0)} = K, \tag{10.45}
$$

where $K \equiv [c_1/(\beta M)]e^{-\alpha_1 t_0}$. Let $r(t) \equiv z(t)/\overline{Z}(t)$ denote a mortgage collection rate that corresponds to the failure rate in reliability theory [1, p. 5]. It is assumed that $r(t)$ is continuous and strictly decreasing because it would be difficult with time to make the collection. Then, we have the following optimum policy:

(i) If $r(0) > K > r(\infty)$, then there exists a finite and unique t^* ($t_0 < t^* < \infty$) that satisfies (10.45).

(ii) If $r(0) \leq K$, then $t^* = t_0$, i.e., the mortgage collection should not be made.

(iii) If $r(\infty) \geq K$, then $t^* = \infty$, i.e., the mortgage collection should be continued until completion.

We had not encountered a number of large-scale bankruptcies until recently in Japan and had not needed to consider the problem of a mortgage collection rate in the bank. Therefore, not enough data have yet been accumulated to estimate a mortgage collection probability $Z(t)$. Suppose for convenience that $Z(t)$ is a Weibull distribution with shape parameter m, i.e., $Z(t) = 1 - e^{-\lambda t^m}$

and $r(t) = \lambda m t^{m-1}$ $(0 < m < 1)$. Then, because $r(t)$ decreases strictly from infinity to zero, from (10.45),

$$t^* - t_0 = \left(\frac{\beta M \lambda m}{c_1} e^{\alpha_1 t_0}\right)^{1/(1-m)}. \tag{10.46}$$

In this case, if $t^* < T$, then the collected capital will be worked again, and conversely, if $t^* > T$, then the capital will be raised newly. In both cases, the total amount $Q(T \mid t_0)$ at time T, when the enterprise has become bankrupt at time t_0, is

$$Q(T \mid t_0) = Q(t^* \mid t_0)e^{\alpha_2(T-t^*)}. \tag{10.47}$$

(3) Expected Earning in Case of Bankruptcy

When the enterprise has become bankrupt at time t_0, the expected earning of the bank at time T is, from (10.42) and (10.47),

$$
\begin{aligned}
P_2(T \mid t_0) &\equiv Q(T \mid t_0) - M e^{\alpha_2 T} \\
&= \left[\beta M e^{\alpha_1 t_0} Z(t^* - t_0) - c_1 \int_0^{t^*-t_0} \overline{Z}(u)\,\mathrm{d}u\right] e^{\alpha_2(T-t^*)} - M e^{\alpha_2 T}.
\end{aligned}
\tag{10.48}
$$

Letting $P_0(T)$ be the expected earning of the bank at time T when a financed period is T,

$$P_0(T) = P_1(T)\overline{F}(T) + \int_0^T P_2(T \mid t_0)\,\mathrm{d}F(t_0). \tag{10.49}$$

We may seek a loan interest rate α_1 that satisfies $P_0(T) \geq 0$ when both distributions of $Z(t)$ and $F(t)$ are given.

Example 10.2.   Suppose that the bankruptcy probability distribution $F(t)$ is discrete. It is assumed that T is divided equally into n, *i.e.*, $T_1 \equiv T/n$, and when the enterprise has become bankrupt during $[(k-1)T_1, kT_1]$, it becomes bankrupt at time kT_1. Then, the distribution $F(t)$ is rewritten as

$$F(kT_1) = \sum_{j=1}^k p_j \quad (k = 1, 2, \ldots, n). \tag{10.50}$$

Thus, the expected earning $P_0(T)$ in (10.49) is

$$P_0(T) = P_1(T)\overline{F}(T) + \sum_{k=1}^n P_2(T \mid kT_1)p_k, \tag{10.51}$$

where $T \equiv nT_1$.

Table 10.5. Loan interest rates for bankruptcy probability p

$\alpha_1(\%)$	3.36	4.08	4.92	5.64	6.36	7.08	7.80	8.40
$p \times 10^3$	0.6	0.8	1.0	1.2	1.4	1.6	1.8	2.0

In general, a bankruptcy probability would be affected greatly by business fluctuations. For example, the number of bankruptcies is small when business is good, is large when it is bad, and becomes constant when it is stable. This probability is also constant for a short time, except for the influence of economic prospects. Thus, when T is 12 months, we suppose that $p_1 = p_2 = \cdots = p_{12} = p$. From the average bankruptcy probability of one year from 1991 to 1992 of enterprises with ranking points 40–60 marked by TEIKOKU DATABANK in Japan, we can consider that the enterprises of such points are normal, and the average bankruptcy probability is about 0.01945. Hence, we set $p \equiv 0.0016 \approx 0.01945/12$.

Next, we show the mortgage collection probability: We have not yet made a statistical investigation of mortgage collection data in Japan, and so, consider that they are similar to those in America. We give the mortgage collection probability, using the data of personal loans at Bank of America [209]. Applying Fig. 5 of [209] to a Weibull distribution in Fig. 10.6, we can estimate the parameters as $\lambda = 0.3$ and $m = 0.24$, i.e., $Z(t) = 1 - \exp(-0.3t^{0.24})$. Furthermore, suppose that a mortgage is set on all amount of loans, the deposit cost is 1.8% of its amount, and the mortgage collection cost is given by the ratio of a clerical cost to the general amount of loans, i.e., $\beta = 1.0$, $\alpha_2 = 1.8\%$, and $c_1 = M/250$. Note that bankruptcy probabilities change a little with each hierarchy of ranking points 40–60. Thus, we compute loan interest rates α_1 that satisfy $P_0(T) = 0$ in (10.51) for several bankruptcy probabilities in Table 10.5.

Furthermore, to use the relation between bankruptcy probability p and loan interest α_1 easily on a business, we use regression analysis and compute in the relation. Figure 10.7 indicates that the values of α_1 increase linearly. Thus, using regression analysis, from Table 10.5,

$$\alpha_1 = 36.36p + 0.012. \qquad (10.52)$$

We could explain in this case that the regression coefficient 36.36 shows the total probabilities of cumulative bankruptcies for three years and the constant value 0.012 shows the expenditure to collect deposits. ∎

10.4 CRL Issue in PKI Architecture

In the *Public Key Infrastructure* (PKI) architecture, the Certificate Management component allows users, administrators, and other principals to the request certification of public keys and revocation of previously certified keys.

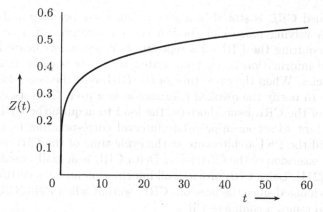

Fig. 10.6. Mortgage collection probability $Z(t)$ for a Weibull distribution $[1 - \exp(-0.3t^{0.24})]$

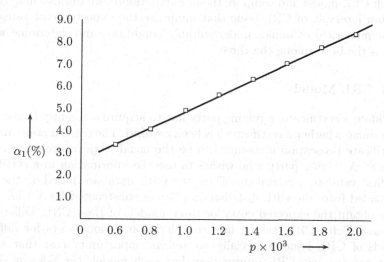

Fig. 10.7. Loan interest α_1 for bankruptcy probability p

When a certificate is issued, it is expected to be in use for its entire validity period. However, various circumstances may cause a certificate to become invalid prior to the expiration of its validity period. Such circumstances involve changes of name and association between subject and *Certification Authority* (CA), and compromises or suspected compromises of the corresponding private key. Under such circumstances, the CA needs to revoke a certificate and periodically issues a signed data structure called *Certificate Revocation List* (CRL) [210]. The $X.509$ defines one method of certificate revocation.

The issued CRL is stored in a server called a *repository* and is open to a public. A relying party can confirm the effectiveness of a certificate by regularly acquiring the CRL of a repository. When a certificate has lapsed, the revoked information is not transmitted to a user because it is issued at planned cycles. When the cycle time of the CRL issue becomes long, it takes a long time to notify the revoked information of a user. Conversely, when the cycle time of the CRL issue shortens, the load to acquire the CRL increases. It is important to set an appropriate interval corresponding to the security business and the PKI architecture at the cycle time of the CRL issue.

As one extension of the CRL issue, Delta CRL is actually used in PKI architecture [211]. Delta CRL provides all information about a certificate whose status has changed since the previous CRL, so that when Delta CRL is issued, the CA also issues a complete CRL.

This section presents three stochastic models of Base CRL, Differential CRL, and Delta CRL, each of which has different types of CRL issues. Introducing various kinds of costs for CRL issues, we obtain the expected costs for each CRL model and compare them. Furthermore, we discuss analytically optimum intervals of CRL issue that minimize the expected cost rates. We present numerical examples under suitable conditions and determine which model is the best among the three.

10.4.1 CRL Models

To validate a certificate, a relying party has to acquire a recently issued CRL to determine whether a certificate has been revoked. The confirmation method of certificate revocation is assumed to be the usual retrieval from the recent CR issue. A relying party who wishes to use the information in a certificate must first validate a certificate. Thus, the CRL database based on the data downloaded from the CRL distribution point is constructed for a user.

We obtain the expected costs for three models of Base CRL, Differential CRL, and Delta CRL, taking into consideration various costs for different methods of CRL issue. Especially, we set an opportunity cost that a user cannot acquire new CRL information. For each model, the CA can decide optimum issue intervals for Base CRL that minimize the expected database construction costs.

We use the following notations when a certificate is revoked in every period k $(k = 1, 2, \ldots, T)$ such as a day:

M_0 : The number of all certificates that have been revoked in Base CRL. M_0 is the total number of CRLs.

T : Interval period between Base CRLs $(T = 1, 2, \ldots)$.

n_k : Expected number of certificates that have been revoked in period k and n_k increases with k $(k = 1, 2, \ldots, T)$.

c_1 : Downloading and communication costs per certificate.

c_2 : File handling cost per downloading Differential CRL or Delta CRL.

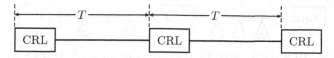

Fig. 10.8. Base CRL that is downloaded once at the beginning of period T

c_3 : Opportunity cost per time when a user cannot acquire new CRL information.

For example, let $F(k)$ be the probability distribution that a certificate is revoked until period k, where $F(0) \equiv 0$. Then, the expected number n_k is given by $n_k = M_0[F(k) - F(k-1)]$ $(k = 1, 2, \ldots, T)$.

(1) Base CRL

Even if a new revoked certification occurs after Base CRL issue, Base CRL is not issued for period T $(T = 1, 2, \ldots)$, *i.e.*, Base CRL is issued only at T intervals (Fig. 10.8). Users download Base CRL once and construct the revoked certificate of CRL database for themselves. There is a possibility that an opportunity cost c_3 may occur when a user cannot acquire new information to the next Base CRL and is assumed to be proportional for the interval of remaining periods. Thus, the expected cost for period T is given by summing the downloading and opportunity costs as follows:

$$C_1(T) = c_1 M_0 + c_3 \sum_{k=1}^{T}(T-k)n_k \quad (T = 1, 2, \ldots). \tag{10.53}$$

We find an optimum T_1^* that minimizes the expected cost rate

$$\widetilde{C}_1(T) \equiv \frac{C_1(T)}{T} = \frac{1}{T}\left[c_1 M_0 + c_3 \sum_{k=1}^{T}(T-k)n_k\right] \quad (T = 1, 2, \ldots). \tag{10.54}$$

It can be seen that $\lim_{T \to \infty} \widetilde{C}_1(T) = \infty$ because n_k increases. Thus, there exists a finite T_1^* $(1 \le T^* < \infty)$. From the inequality $\widetilde{C}_1(T+1) - \widetilde{C}_1(T) \ge 0$,

$$\sum_{k=1}^{T} kn_k \ge \frac{c_1 M_0}{c_3}. \tag{10.55}$$

Therefore, there exists a finite and unique minimum T_1^* $(1 \le T_1^* < \infty)$ that satisfies (10.55), and

$$c_3 \sum_{k=1}^{T_1^*} n_k \le \frac{C_1(T_1^*)}{T_1^*} < c_3 \sum_{k=1}^{T_1^*+1} n_k. \tag{10.56}$$

Fig. 10.9. Differential CRL where Δ means Differential CRL

In particular, when $n_k = n$, (10.55) becomes

$$\frac{T(T+1)}{2} \geq \frac{c_1 M_0}{c_3 n},\qquad(10.57)$$

that corresponds to the type of inequality (3.5) in Chap. 3.

(2) Differential CRL

Differential CRL is continuously issued for every period k $(k = 1, 2, \ldots, T)$ after Base CRL issue (Fig. 10.9). The number of revoked certificates in Differential CRL is the total of newly revoked from the previous Differential CRL to this one. The full CRL database for a user is constructed from the previous Base CRL and is updated by every Differential CRL. Thus, a handling cost c_2 is needed for the frequencies of downloaded Differential CRL files, *i.e.*, this cost increases with the frequency of downloaded CRL.

It is assumed that a user is not affected the opportunity cost c_3 because a user can acquire the revoked information by Differential CRL issue in a short period. The expected cost for period T is the total of downloading costs for Base CRL and Differential CRL, and the handling cost for the number of Differential CRLs filed as follows:

$$C_2(T) = c_1 M_0 + c_1 \sum_{k=1}^{T} n_k + c_2 \sum_{k=1}^{T} k \qquad (T = 1, 2, \ldots).\qquad(10.58)$$

The description of the method of Differential CRL is not shown in $X.\,509$. However, Differential CRL exports only files that have changed since the last Differential CRL or Base CRL and imports files of all Differential CRL and the last Base CRL. The reason that the generation of CRLs per time increases in proportion to its amount is that if the registration number of Base CRL increases, Differential CRL would be efficient.

We find an optimum T_2^* that minimizes the expected cost rate

$$\tilde{C}_2(T) \equiv \frac{C_2(T)}{T} = \frac{1}{T}\left(c_1 M_0 + c_1 \sum_{k=1}^{T} n_k + c_2 \sum_{k=1}^{T} k\right) \qquad (T = 1, 2, \ldots).$$
$$(10.59)$$

There exists a finite T_2^* $(1 \leq T_2^* < \infty)$ because $\lim_{T \to \infty} \tilde{C}_2(T) = \infty$. From the inequality $\tilde{C}_2(T+1) - \tilde{C}_2(T) \geq 0$,

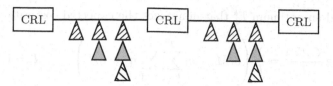

Fig. 10.10. Delta CRL where Δ means Differential CRL

$$c_1 \sum_{k=1}^{T}(n_{T+1} - n_k) + c_2 \sum_{k=1}^{T} k \geq c_1 M_0 \qquad (T = 1, 2, \ldots). \tag{10.60}$$

Therefore, there exists a finite and unique minimum T_2^* $(1 \leq T_2^* < \infty)$ that satisfies (10.60) because the left-hand side of (10.60) increases strictly with T, and

$$c_1 n_{T_2^*} + c_2 T_2^* \leq \frac{C_2(T_2^*)}{T_2^*} < c_1 n_{T_2^*+1} + c_2(T_2^* + 1). \tag{10.61}$$

In particular, when $n_k = n$, (10.60) becomes

$$\frac{T(T+1)}{2} \geq \frac{c_1 M_0}{c_2}. \tag{10.62}$$

(3) Delta CRL

Delta CRL is continuously issued for every period k $(k = 1, 2, \ldots, T)$ after Base CRL issue (Fig. 10.10). Delta CRL is a small CRL that provides information about certificates whose status changed since the previous Base CRL [211], *i.e.*, the number of revoked certificates in Delta CRL is the total of accumulated revoked certificates from the previous Base CRL issue. The full CRL database for a user is constructed from Base CRL and the previous Delta CRL.

It is assumed that an opportunity cost c_3 is not generated because a user can acquire the revoked information by Delta CRL issue in a short period. The expected cost for period T is the total of downloading costs for Base CRL and Delta CRL and the handling cost of files as follows:

$$C_3(T) = c_1 M_0 + c_1 \sum_{k=1}^{T} \sum_{j=1}^{k} n_j + c_2 T \qquad (T = 1, 2, \ldots). \tag{10.63}$$

The method of operating Delta CRL is introduced in $X.509$ and has the advantage that the full CRL can always be done any period. A user, who needs more up-to-date certificate status obtained the previous CRL issue, can download the latest Delta CRL. This tends to be significantly smaller than the full CRL, would reduce the load in the repository, and improve the response time for a user [212].

We find an optimum T_3^* that minimizes the expected cost rate

$$\tilde{C}_3(T) \equiv \frac{C_3(T)}{T} = \frac{1}{T}\left(c_1 M_0 + c_1 \sum_{k=1}^{T}\sum_{j=1}^{k} n_j + c_2 T\right) \qquad (T = 1, 2, \dots).$$

(10.64)

There exists a finite T_3^* $(1 \le T_3^* < \infty)$ because $\lim_{T\to\infty} C_3(T)/T = \infty$. From the inequality $\tilde{C}_3(T+1) - \tilde{C}_3(T) \ge 0$,

$$T \sum_{k=1}^{T+1} n_k - \sum_{k=1}^{T}\sum_{j=1}^{k} n_j \ge M_0 \qquad (T = 1, 2, \dots).$$

(10.65)

There exists a finite and unique minimum T_3^* $(1 \le T_3^* < \infty)$ that satisfies (10.65) because the left-hand side of (10.65) increases strictly with T, and

$$c_1 \sum_{k=1}^{T_3^*} n_k \le \frac{C_3(T_3^*)}{T_3^*} - c_2 < c_1 \sum_{k=1}^{T_3^*+1} n_k.$$

(10.66)

In particular, when $n_k = n$, (10.65) becomes

$$\frac{T(T+1)}{2} \ge \frac{M_0}{n}.$$

(10.67)

When $c_1/c_2 = 1/n$ and $c_1 = c_3$, i.e., $nc_1 = c_2$ and $c_1 = c_3$, $T_1^* = T_2^* = T_3^*$.

(4) Comparisons of Expected Costs

We compare the expected costs $C_1(T)$ in (10.53), $C_2(T)$ in (10.58), and $C_3(T)$ in (10.63) for a specified T. For $T = 1, C_2(1) = C_3(1) > C_1(1)$.

The following three relations among the expected costs are obtained:

$$C_1(T) \ge C_2(T) \Leftrightarrow \frac{\sum_{k=1}^{T}[c_3(T-k) - c_1]n_k}{\sum_{k=1}^{T} k} \ge c_2,$$

(10.68)

$$C_1(T) \ge C_3(T) \Leftrightarrow \frac{\sum_{k=1}^{T}[c_3(T-k) - c_1(T-k+1)]n_k}{T} \ge c_2,$$

(10.69)

$$C_3(T) \ge C_2(T) \Leftrightarrow \frac{c_1 \sum_{k=1}^{T-1}(T-k)n_k}{\sum_{k=1}^{T-1} k} \ge c_2.$$

(10.70)

Therefore, from (10.68) and (10.69), when

$$c_2 \ge \frac{\sum_{k=1}^{T}[c_3(T-k) - c_1]n_k}{\sum_{k=1}^{T} k}, \qquad c_2 \ge \frac{\sum_{k=1}^{T}[c_3(T-k) - c_1(T-k+1)]n_k}{T},$$

(10.71)

the expected cost $C_1(T)$ is minimum. From (10.68) and (10.70), when

$$\frac{\sum_{k=1}^{T}[c_3(T-k)-c_1]n_k}{\sum_{k=1}^{T}k} \geq c_2, \qquad \frac{c_1\sum_{k=1}^{T-1}(T-k)n_k}{\sum_{k=1}^{T-1}k} \geq c_2, \qquad (10.72)$$

the expected cost $C_2(T)$ is minimum. From (10.69) and (10.70), when

$$\frac{\sum_{k=1}^{T}[c_3(T-k)-c_1(T-k+1)]n_k}{T} \geq c_2 \geq \frac{c_1\sum_{k=1}^{T-1}(T-k)n_k}{\sum_{k=1}^{T-1}k}, \qquad (10.73)$$

the expected cost $C_3(T)$ is minimum.

Next, suppose that $n_k = n$. Then, from (10.71), when

$$\frac{c_2}{n} \geq \frac{c_3(T-1)-2c_1}{T+1}, \qquad \frac{c_2}{n} \geq c_3\frac{T-1}{2} - c_1\frac{T+1}{2},$$

$C_1(T)$ is minimum. From (10.72), when

$$\frac{c_3(T-1)-2c_1}{T+1} \geq \frac{c_2}{n}, \qquad c_1 \geq \frac{c_2}{n},$$

$C_2(T)$ is minimum. From (10.73), when

$$c_3\frac{T-1}{2} - c_1\frac{T+1}{2} \geq \frac{c_2}{n} \geq c_1,$$

$C_3(T)$ is minimum. The above results indicate that $C_1(T)$ decreases when c_2 increases. Similarly, $C_2(T)$ decreases when both c_1 and c_3 increase, and $C_3(T)$ decreases when c_3 increases but c_1 decreases.

Example 10.3.   The revocation has occurred daily almost equally and its number of certificates is constant, *i.e.*, $n \equiv n_k$. When $M_0 = 10,000$ and $n = 40$, we present the optimum interval T_1^* (days) and $C_1(T_1^*)/(c_1T_1^*)$ in Table 10.6. This indicates that $T_1^* = 1$ for $c_3/c_1 \geq 250.0$, *i.e.*, we should issue the CRL every day. Clearly, the optimum interval increases when the ratio of cost c_3/c_1 decreases. For example, when $c_3/c_1 = 16.7$, T_1^* is 5 days and $C_1(T_1^*)/(c_1T_1^*) = 3,336$. Similarly, we present the optimum interval T_2^* (days) and $C_2(T_2^*)/(c_1T_2^*)$ in Table 10.7. This indicates that if c_2/c_1 is very large, we should issue the CRL every day. However, because c_2/c_1 is the ratio of the initial construction cost of the database to its additional handling cost for a user, it would be less than about 21.5. In this case, Base CRL should be done within a month, while Differential CRL would be issued every day. Finally, the optimum interval is $T_3^* = 22$ (days) from (10.67), regardless of cost c_i ($i = 1,2,3$). When $T_3^* = 22$, we show $C_3(T_3^*)/(c_1T_3^*)$ for c_2/c_1 in Table 10.8. Comparing Tables 10.7 and 10.8, the expected cost rates in Table 10.8 are smaller than those in Table 10.7 for $c_2/c_1 \geq 47.6$. If the number n of certificates becomes larger, the optimum interval T_3^* becomes shorter from (10.67). Thus, comparing the optimum intervals T_2^* and T_3^*, if n becomes larger, T_3^* becomes shorter, and Differential CRL is more effective than Delta CRL. Conversely, if c_2/c_1 becomes larger, T_2^* becomes shorter, and Delta CRL improves more than Differential CRL. ∎

Table 10.6. Optimum interval T_1^* for Model 1 when $M_0 = 10,000$ and $n = 40$

T_1^*	c_3/c_1	$C_1(T_1^*)/(c_1 T_1^*)$
1	250.0	10000
5	16.7	3336
10	4.5	1810
15	2.1	1255
20	1.2	956
25	0.8	784
30	0.5	623

Table 10.7. Optimum interval T_2^* for Model 2 when $M_0 = 10,000$ and $n = 40$

T_2^*	c_2/c_1	$C_2(T_2^*)/(c_1 T_2^*)$
1	10000	20040
5	666.7	4040
10	181.8	2040
15	83.3	1373
20	47.6	1040
22	39.5	954
25	30.8	840
30	21.5	707

Table 10.8. Optimum interval T_3^* for Model 3 when $M_0 = 10,000$ and $n = 40$

T_3^*	c_2/c_1	$C_3(T_3^*)/(c_1 T_3^*)$
	10000	10915
	666.7	1581
	181.8	1096
22	83.3	998
	47.6	962
	39.5	949
	30.8	945
	21.5	936

References

1. Nakagawa T (2005) Maintenance Theory of Reliability. Springer, London.
2. Barlow RE, Proschan F (1965) Mathematical Theory of Reliability. Wiley, New York.
3. Nakagawa T (2007) Shock and Damage Models in Reliability Theory. Springer, London.
4. Wang H, Pham H (2007) Reliability and Optimal Maintenance. Springer, London.
5. Ushakov IA (1994) Handbook of Reliability Engineering. Wiley, New York.
6. Birolini A (1999) Reliability Engineering Theory and Practice. Springer, New York.
7. Kuo W, Prsad VR, Tillman FA, Hwang CL (2001) Optimal Reliability Design. Cambridge University Press, Cambridge.
8. Sung CS, Cho YK, Song SH (2003) Combinatorial reliability optimization. In: Pham H (ed) Handbook of Reliability Engineering. Springer, London: 91–114.
9. Levitin G (2007) Computational Intelligence in Reliability Engineering. Springer, Berlin.
10. Abd-El-Barr M (2007) Reliable and Fault-Tolerant. Imperial College Press, London.
11. Lee PA, Anderson T (1990) Fault Tolerance - Principles and Practice. Springer, Wien.
12. Lala PK (1985) Fault Tolerant and Fault Testable Hardware Design. Prentice-Hall, London.
13. Nanya T (1991) Fault Tolerant Computer. Ohm, Tokyo.
14. Gelenbe E (2000) System Performance Evaluation. CRC, Boca Raton, FL.
15. Karlin S, Taylor HM (1975) A First Course in Stochastic Processes. Academic Press, New York.
16. Çinlar E (1975) Introduction to Stochastic Processes. Prentice-Hall, Englewood Cliffs, NJ.
17. Osaki S (1992) Applied Stochastic System Modelling. Springer, Berlin.
18. Satow T, Yasui K, Nakagawa T (1996) Optimal garbage collection policies for a database in a computer system. RAIRO Oper Res 30: 359–372.
19. Ito K, Nakagawa T (1992) Optimal inspection policies for a system in storage. Comput Math Appl 24: 87–90.

236   References

20. Ito K, Nakagawa T (2004) Comparison of cyclic and delayed maintenance for a phased array radar. J Oper Res Soc Jpn 47: 51–61.
21. Ito K, Nakagawa T (2003) Optimal self-diagnosis policy for FADEC of gas turbine engines. Math Comput Model 38: 1243–1248.
22. Ito K, Nakagawa T (2006) Maintenance of a cumulative damage model and its application to gas turbine engine of co-generation system. In: Pham H (ed) Reliability Modelling Analysis and Optimization. World Scientific, Singapore: 429–438.
23. Cox JC, Rubinstein M (1985) Options Markets. Prentice-Hall, Englewood Cliffs, NJ.
24. Beichelt F (2006) Stochastic Processes in Science, Engineering and Finance. Chapman & Hall, Boca Raton FL.
25. Shannon CE, Weaver W (1949) The Mathematical Theory of Communication. University of Illinois, Chicago.
26. Kunisawa K (1975) Entropy Models. Nikka Giren Shuppan, Tokyo.
27. Pham H (2003) Reliability of systems with multiple failure modes. In: Pham H (ed) Handbook of Reliability Engineering. Springer, London: 19–36.
28. Blokus A (2006) Reliability analysis of large systems with dependent components. Inter J Reliab Qual Saf Eng 13: 1–14.
29. Yasui K, Nakagawa T, Osaki S (1988) A summary of optimum replacement policies for a parallel redundant system. Microelectron Reliab 28: 635–641.
30. Nakagawa T, Yasui K (2005) Note on optimal redundant policies for reliability models. J Qual Maint Eng 11: 82–96.
31. Zuo MJ, Huang J, Kuo W (2003) Multi-state k-out-of-n systems. In: Pham H (ed) Handbook of Reliability Engineering. Springer, London: 3–17.
32. Nakagawa T, Qian CH (2002) Note on reliabilities of series-parallel and parallel-series systems. J Qual Maint Eng 8: 274–280.
33. Yasui K, Nakagawa T, Sandoh H (2002) Reliability models in data communication systems. In: Osaki S (ed) Stochastic Models in Reliability and Maintenance. Springer, Berlin: 281–301.
34. Nakagawa T (1984) Optimal number of units for a parallel system. J Appl Probab 21: 431–436.
35. Nakagawa T (1984) A summary of discrete replacement policies. Euro J Oper Res 17: 382–392.
36. Linton DG, Saw JG (1974) Reliability analysis of the k-out-of-n: F system. IEEE Trans Reliab R-23: 97–103.
37. Nakagawa T (1985) Optimization problems in k-out-of-n systems. IEEE Trans Reliab R-34: 248–250.
38. Kenyon RL, Newell RL (1983) Steady-state availability of k-out-of-n: G system with single repair. IEEE Trans Reliab R-32: 188–190.
39. Chang GJ, Cui L, Hwang FK (2000) Reliability of Consecutive-k Systems. Kluwer, Dordrecht.
40. Nakagawa T (1986) Modified discrete preventive maintenance policies. Nav Res Logist Q 33: 703–715.
41. Lin S, Costello DJ Jr, Miller MJ (1984) Automatic-repeat-request error-control scheme. IEEE Trans Commun Mag 22: 5–17.
42. Moeneclaey M, Bruneel H, Bruyland I, Chung DY (1986) Throughput optimization for a generalized stop-and-wait ARQ scheme. IEEE Trans Commun COM-34: 205–207.

43. Fantacci R (1990) Performation evaluation of efficient continuous ARQ protocols. IEEE Trans Commun 38: 773–781.

44. Yasui K, Nakagawa T (1992) Reliability consideration on error control policies for a data communication system. Comput Math Appl 24: 51–55.

45. Koike S, Nakagawa T, Yasui K (1995) Optimal block length for basic mode data transmission control procedure. Math Comput Model 22: 167–171.

46. Saaty TL (1961) Elements of Queueing Theory with Applications. McGraw-Hill, New York.

47. Kuo W, Prasad VR, Tillman FA, Hwang CL (2001) Optimal Reliability Design. Cambridge University Press, Cambridge.

48. Nakagawa T, Yasui K, Sando H (2004) Note on optimal partition problems in reliability models. J Qual Maint Eng 10: 282–287.

49. Nakagawa T, Mizutani S (2007) A summary of maintenance policies for a finite interval. Reliab Eng Syst Saf.

50. Sandoh H, Kawai H (1991) An optimal N-job backup policy maximizing availability for a hard computer disk. J Oper Res Soc Jpn 34: 383–390.

51. Sandoh H, Kawai H (1992) An optimal $1/N$ backup policy for data floppy disks under efficiency basis. J Oper Res Soc Jpn 35: 366–372.

52. Conffman EG Jr, Gilbert EN (1990) Optimal strategies for scheduling checkpoints and preventive maintenance. IEEE Trans Reliab R-39: 9–18.

53. Steele GL Jr (1975) Multiprocessing compactifying garbage collection. Communications ATM 18: 495–508.

54. Satow T, Yasui K, Nakagawa T (1996) Optimal garbage collection policies for a database in a computer system. RAIRO Oper Res 30: 359–372.

55. Yoo YB, Deo N (1998) A comparison of algorithms for terminal-pair reliability. IEEE Trans Reliab R-37: 210–215.

56. Ke WJ, Wang SD (1997) Reliability evaluation for distributed computing networks with imperfect nodes. IEEE Trans Reliab 46: 342–349.

57. Imaizumi M, Yasui K, Nakagawa T (2003) Reliability of a job execution process using signatures. Math Comput Model 38: 1219–1223.

58. Fukumoto S, Kaio N, Osaki S (1992) A study of checkpoint generations for a database recovery mechanism. Comput Math Appl 24: 63–70.

59. Vaidya NH (1998) A case for two-level recovery schemes. IEEE Trans Comput 47: 656–666.

60. Ziv A, Bruck J (1997) Performance optimization of checkpointing schemes with task duplication. IEEE Trans Comput 46: 1381–1386.

61. Ziv A, Bruck J (1998) Analysis of checkpointing schemes with task duplication. IEEE Trans Comput 47: 222–227.

62. Nakagawa S, Fukumoto S, Ishii N (2003) Optimal checkpointing intervals for a double modular redundancy with signatures. Comput Math Appl 46: 1089–1094.

63. Mahmood A, McCluskey EJ (1988) Concurrent error detection using watchdog processors – A survey. IEEE Trans Comput 37: 160–174.

64. Imaizumi M, Yasui K, Nakagawa T (1998) Reliability analysis of microprocessor systems with watchdog processors. J Qual Maint Eng 4: 263–272.

WP, Voelker JA (1976) A survey of maintenance models: The control lance of deteriorating systems. Nav Res Logist Q 23: 353–388.

Smith ML (1981) Optimal maintenance models for systems subject A review. Nav Res Logist Q 28:47–74.

67. Thomas LC (1986) A survey of maintenance and replacement models for maintainability and reliability of multi-item systems. Reliab Eng 16: 297–309.

68. Valdez-Flores C, Feldman RM (1989) A survey of preventive maintenance models for stochastic deteriorating single-unit systems. Nav Logist Q 36: 419–446.

69. Jensen U (1991) Stochastic models of reliability and maintenance: An overview. In: Özekici S (ed) Reliability and Maintenance of Complex Systems. Springer, Berlin: 3–6.

70. Gertsbakh I (2000) Reliability Theory with Applications to Preventive Maintenance. Springer, Berlin.

71. Ben-Daya M, Duffuaa SO, Raouf A (2000) Overview of maintenance modeling areas. In: Ben-Daya M, Duffuaa SO, Raouf A (eds) Maintenance, Modeling and Optimization. Kluwer Academic, Boston: 3–35.

72. Osaki S (ed) (2002) Stochastic Models in Reliability and Maintenance. Springer, Berlin.

73. Pham H (ed) (2003) Handbook of Reliability Engineering. Springer, London.

74. Hudson WR, Haas R, Uddin W (1997) Infrastructure Management. McGraw-Hill, New York.

75. Lugtigheid D, Jardine AKS, Jiang X (2007) Optimizing the performance of a repairable system under a maintenance and repair contract. Qual Reliab Eng Inter 23: 943–960.

76. Christer AH (2003) Refined asymptotic costs for renewal reward process. J Oper Res Soc 29: 577–583.

77. Ansell J, Bendell A, Humble S (1984) Age replacement under alternative cost criteria. Manage Sci 30:358–367.

78. Hariga M, Al-Fawzan MA (2000) Discounted models for the single machine inspection problem. In: Ben-Daya M, Duffuaa SO, Raouf A (eds) Maintenance, Modeling and Optimization. Kluwer Academic, Boston: 215–243.

79. Nakagawa T (1988) Sequential imperfect preventive maintenance policies. IEEE Trans Reliab 37: 295–298.

80. Nakagawa T (2000) Imperfect preventive maintenance models. In: Ben-Daya M, Duffuaa SO, Raouf A (eds) Maintenance, Modeling and Optimization. Kluwer Academic, Boston: 201–214.

81. Nakagawa T (2002) Imperfect preventive maintenance models. In: Osaki S (ed) Stochastic Models in Reliability and Maintenance. Springer, Berlin: 125–143.

82. Pham H, Wang H (1996) Imperfect maintenance. Eur J Oper Res: 425–438.

83. Wang H, Pham H (2003) Optimal imperfect maintenance models. In: Pham H (ed) Handbook of Reliability Engineering. Springer, London: 397–414.

84. Brown M, Proschan F (1983) Imperfect repair. J Appl Probab 20: 851–859.

85. Vaurio JK (1999) Availability and cost functions for periodically inspected preventively maintained units. Reliab Eng Sys Saf 63: 133–140.

86. Nakagawa T (2003) Maintenance and optimum policy. In: Pham H (ed) Handbook of Reliability Engineering. Springer, London: 367–395.

87. Osaki T, Dohi T, Kaio N (2004) Optimal inspection policies with an equality constraint based on the variational calculus approach. In: Dohi T, Yun WY (eds) Advanced Reliability Modeling. World Scientific, Singapore: 387–394.

88. Keller JB (1974) Optimum checking schedules for systems subject to random failure. Manage Sci 21: 256–260.

89. Kaio N, Osaki S (1989) Comparison of inspection policies. J Oper Res Soc 40: 499–503.

90. Visicolani B (1991) A note on checking schedules with finite horizon. RAIRO Oper Res 25: 203–208.

91. Sobczyk K, Trebick J (1989) Modelling of random fatigue by cumulative jump process. Eng Fracture Mech 34: 477–493.

92. Scarf PA, Wang W, Laycok PJ (1996) A stochastic model of crack growth under periodic inspections. Reliab Eng Syst Saf 51: 331–339.

93. Hopp WJ, Kuo YL (1998) An optimal structured policy for maintenance of partially observable aircraft engine components. Nav Res Logist 45: 335–352.

94. Lukić M, Cremona C (2001) Probabilistic optimization of welded joints maintenance versus fatigue and fracture. Reliab Eng Syst Saf 72: 253–264.

95. Garbotov Y, Soares CG (2001) Cost and reliability based strategies for fatigue maintenance planning of floating structures. Reliab Eng Syst Saf 73: 293–301.

96. Petryna YS, Pfanner D, Shangenberg F, Kraätzig WB (2002) Reliability of reinforced concrete structures under fatigue. Reliab Eng Syst Saf 77: 253–261.

97. Campean IF, Rosala GF, Grove DM, Henshall E (2005) Life modelling of a plastic automotive component. In: Proc. Annal Reliability and Maintainability Symposium: 319–325.

98. Sobczyk K (1987) Stochastic models for fatigue damage of materials. Adv Appl Probab 19: 652–673.

99. Sobczyk K, Spencer BF Jr (1992) Random Fatigue: From Data to Theory. Academic Press, New York.

100. Dasgupta A. Pecht M (1991) Material failure mechanisms and damage models. IEEE Trans Reliab 40: 531–536.

101. Kijima M, Nakagawa T (1992) Replacement policies of a shock model with imperfect preventive maintenance. Eur J Oper Res 57: 100–110.

102. Nakagawa T (1986) Periodic and sequential preventive maintenance policies. J Appl Probab 23: 536–542.

103. Mie J (1995) Bathtub failure rate and upside-down bathtub mean residual life. IEEE Trans Reliab 44: 388–391.

104. Barlow RE, Proschan F (1975) Statistical Theory of Reliability and Life Testing Probability Models. Holt, Rinehart & Winston, New York.

105. Sandoh H, Nakagawa T (2003) How much should we reweigh?. J Oper Res Soc 54: 318–321.

106. Nakagawa S, Ishii N, Fukumoto S (1998) Evaluation measures of archive copies for file recovery mechanism. J Qual Maint Eng 4: 291–298.

107. Nakagawa T (2004) Five further studies of reliability models. In: Dohi T, Yun WY (eds) Advanced Reliability Modeling. Word Scientific, Singapore: 347–361.

108. Nakagawa T, Mizutani S, Sugiura T (2005) Note on the backward time of reliability models. Eleventh ISSAT International Conference on Reliability and Quality in Design: 219–222.

109. Naruse K, Nakagawa S, Okuda Y (2005) Optimal checking time of backup operation for a database system. Proceedings of International Workshop on Recent Advances in Stochastic Operations Research: 179–186.

110. Pinedo M (2002) Scheduling Theory, Algorithms, and Systems. Prentice-Hall, Englewood Cliffs, NJ.

111. Durham SD, Padgett WJ (1990) Estimation for a probabilistic stress-strength model. IEEE Trans Reliab 39: 199–203.

112. Finkelstein MS (2002) On the reversed hazard rate. Reliab Eng Syst Saf 78: 71–75.

113. Block HW, Savits TH, Singh H (1998) The reversed hazard rate function. Probab Eng Inform Sci 12: 69–90.
114. Chandra NK, Roy D (2001) Some results on reversed hazard rate. Probab Eng Inform Sci 15: 95–102.
115. Keilson J, Sumita U (1982) Uniform stochastic ordering and related inequalities. Can J Statis 10: 181–198.
116. Shaked M, Shanthikumar (1994) Stochastic Orders and Their Applications. Academic Press, New York.
117. Gupta RD, Nanda AK (2001) Some results on reversed hazard rate ordering. Common Statist-Theory Meth 30: 2447–2457.
118. Morey RC (1967) A certain for the economic application of imperfect inspection. Oper Res 15: 695–698.
119. Lees M (ed) (2003) Food Authenticity and Traceability. Woodhead, Cambrige.
120. Reuter A (1984) Performance analysis of recovery techniques. ACM Trans. Database Syst 9: 526–559.
121. Fukumoto S, Kaio N, Osaki S (1992) A study of checkpoint generations for a database recovery mechanism. Comput Math Appl 24: 63–70.
122. Sandoh H, Igaki N (2001) Inspection policies for a scale. J Qual Maint Eng 7: 220–231.
123. Sandoh H, Igaki N (2003) Optimal inspection policies for a scale. Comput Math Appl 46: 1119–1127.
124. Sandoh H, Nakagawa T, Koike S (1993) A Bayesian approach to an optimal ARQ number in data transmission. Electro Commun Jpn 76: 67–71.
125. Schwarts M (1987) Telecommunication Networks: Protocols, Modeling and Analysis. Addison Reading, Wesley, MA.
126. Chang JF, Yang TH (1993) Multichannel ARQ protocols. IEEE Trans Commun 41: 592–598.
127. Lu DL, Chang JF (1993) Performance of ARQ protocols in nonindependent channel errors. IEEE Trans Commun 41: 721–730.
128. Falin GI, Templeton JGC (1977) Retrial Queues. Chapman & Hall, London.
129. Nakagawa T, Yasui K (1989) Optimal testing-policies for intermittent faults. IEEE Trans Reliab 38: 577–580.
130. Pyke R (1961) Markov renewal process: Definitions and preliminary properties. Ann Math Statist 32: 1231–1242.
131. Pyke R (1961) Markov renewal process with finitely many states. Ann Math Statist 32: 1243–1259.
132. Marz HF, Waller RA (1992) Bayesian Reliability Analysis. Wiley, New York.
133. Nakagawa T, Yasui K, Sandoh H (1993) An optimal policy for a data transmission system with intermittent faults. Trans Inst Electron Inform Commun Eng J76-A: 1201–1206.
134. Yasui K, Nakagawa T, Sandoh H (1995) An ARQ policy for a data transmission system with three types of error probabilities. Trans Inst Electron Inform Commun Eng J78-A: 824–830.
135. Yasui K, Nakagawa T (1995) Reliability consideration of a selective-repeat ARQ policy for a data communication system. Microelectron Reliab 35: 41–44.
136. Yasui K, Nakagawa T (1997) Reliability analysis of a hybrid ARQ system with finite response time. Trans Inst Electron Inform Commun Eng J80-A: 221–227.

137. Yasui K, Nakagawa T, Imaizumi M (1998) Reliability evaluations of hybrid ARQ policies for a data communication systems. Int J Reliab Qual Saf Eng 5: 15–28.
138. Nakagawa T, Motoori M, Yasui K (1990) Optimal testing policy for a computer system with intermittent faults. Reliab Eng Syst Saf 27: 213–218.
139. Pradhan DK, Vaidya NH (1994) Roll-forward checkpointing scheme: A novel fault-tolerant architecture. IEEE Trans Comput 43: 1163–1174.
140. Ling Y, Mi J, Lin X (2001) A variational calculus approach to optimal checkpoint placement. IEEE Trans Comput 50: 699–707.
141. Pradhan DK, Vaidya NH (1992) Rollforward checkpointing scheme: Concurrent retry with nondedicated spares. IEEE Workshop on Fault-Tolerant Parallel and Distributed Systems: 166–174.
142. Pradhan DK, Vaidya NH (1994) Roll-forward and rollback recovery: Performance-reliability trade-off. In: 24th Int Symp on Fault-Tolerant Comput: 186–195.
143. Kim H, Shin KG (1996) Design and analysis of an optimal instruction-retry policy for TMR controller computers. IEEE Trans Comput 45: 1217–1225.
144. Ohara M, Suzuki R, Arai M, Fukumoto S, Iwasaki K (2006) Analytical model on hybrid state saving with a limited number of checkpoints and bound rollbacks. IEICE Trans Fundam E89-A: 2386–2395.
145. Nakagawa S, Fukumoto S, Ishii N (1998) Optimal checkpoint interval for redundant error detection and masking systems. First Euro-Japanese Workshop on Stochastic Risk Modeling for Finance, Insurance, Production and Reliability 2.
146. Naruse K, Nakagawa T, Maeji S (2006) Optimal checkpoint intervals for error detection by multiple modular redundancies. Advanced Reliability Modeling II: Reliability Testing and Improvement (AIWARM2006): 293–300.
147. Naruse K, Nakagawa T, Maeji S (2007) Optimal sequential checkpoint intervals for error detection. Proceedings of International Workshop on Recent Advances in Stochastic Operations Research II (2007 RASOR Nanzan): 185–191.
148. Nakagawa S, Okuda Y, Yamada S (2003) Optimal checkpointing interval for task duplication with spare processing. Ninth ISSAT International Conference on Reliability and Quality in Design: 215–219.
149. Nakagawa S, Fukumoto S, Ishii N (2003) Optimal checkpointing intervals of three error detection schemes by a double modular redundancy. Math Comput Model 38: 1357–1363.
150. Ben-Daya M, Duffuaa SO, Raouf A (eds) (2000) Maintenance, Modelling and Optimization. Kluwer Academic, Boston.
151. Nakagawa T (1985) Continuous and discrete age-replacement policies. J Oper Res Soc 36: 147–154.
152. Sugiura T, Mizutani S, Nakagawa T (2004) Optimal random replacement policies. In: Tenth ISSAT International Conference on Reliability and Quality in Design: 99–103.
153. Nakagawa T (1981) Generalized models for determining optimal number of minimal repairs before replacement. J Oper Res Jpn 24: 325–337.
154. Nakagawa T (1983) Optimal number of failures before replacement time. IEEE Trans Reliab R-32: 115–116.
155. Nakagawa T (1984) Optimal policy of continuous and discrete replacement with minimal repair at failure. Nav Res Logist Q 31: 543–550.

242   References

156. Wu S, Croome DC (2005) Preventive maintenance models with random maintenance quality. Reliab Eng Syst Saf 90: 99–105.
157. Duchesne T, Lawless JF (2000) Alternative times scales and failure time models. Life Time Data Anal 6: 157–179.
158. Yun WY, Choi CH (2000) Optimum replacement intervals with random time horizon. J Qual Maint Eng 6: 269–274.
159. Ito K, Nakagawa T (1992) An optimal inspection policy for a storage system with finite number of inspections. J Reliab Eng Assoc Jpn 19: 390–396.
160. Ito K, Nakagawa T (1995) Extended optimal inspection policies for a system in storage. Math Comput Model 22: 83–87.
161. Ito K, Nakagawa T (2000) Optimal inspection policies for a storage system with degradation at periodic test. Math Comput Model 31: 191–195.
162. Waldrop MM (1992) The Emerging Science at the Edge of Order and Chaos. Sterling Lord Literistic, New York.
163. Badi R, Polti A (1997) Complexity: Hierarchical Structures and Scaling in Physics. Cambridge University Press, Cambridge.
164. Lala PK (2001) Self-Checking and Fault-Tolerant Digital Design. Morgan Kaufmann, San Francisco.
165. Pukite J, Pukite P (1998) Modeling for Reliability Analysis. Inst Electric Electron Eng, New York.
166. Nakagawa T, Yasui K (2003) Note on reliability of a system complexity. Math Comput Model 38: 1365-1371.
167. Shannon CE, Weaver W (1949) The Mathematical Theory of Communication. University of Illinois, Chicago.
168. Kullback S (1958) Information Theory and Statistics. Wiley, New York.
169. Ash R (1965) Information Theory. Wiley, New York.
170. Kunisawa K (1975) Entropy Models. Nikka Giren Shuppan, Tokyo.
171. Miller Jr JE, Kulp RW, Orr GE (1984) Adaptive probability distribution estimation based upon maximum entropy. IEEE Trans Reliab R-33: 353–357.
172. Teitler S, Rajagopal AK, Ngai KL (1986) Maximum entropy and reliability distributions. IEEE Trans Reliab R-35: 391–395.
173. Ohi F, Suzuki T (2000) Entropy and safety monitoring systems. Jpn J Ind Appl Math 17: 59–71.
174. Billingsley P (1960) Ergodic Theory and Information. Wiley, New York.
175. McCabe TJ (1976) A complexity measure. IEEE Trans Software Eng SE-2: 308–320.
176. Weyuker EJ (1988) Evaluating software complexity measures. IEEE Trans Software Eng 14: 1357–1365.
177. Davis JS, LeBlank RJ (1988) A study of the applicability of complexity measures. IEEE Trans Software Eng 14: 1366–1372.
178. Meitzler T, Gerhard G, Singh H (1996) On modification of the relative complexity metric. Microelectron Reliab 36: 469–475.
179. Park J, Jung W, Ha J (2001) Development of the step complexity measure for emergency operating produres using entropy concepts. Reliab Eng Syst Saf 71: 115–130.
180. Nakagawa T, Yasui K (2003) Note on reliability of a system complexity considering entropy. J Qual Maint 9: 83–91.
181. Ball MO (1986) Computational complexity of network reliability analysis: An overview. IEEE Trans Reliab R-35: 230–239.

182. Yoo YB, Deo N (1988) A comparison of algorithms for terminal-pair reliability. IEEE Trans Reliab 37: 210–215.

183. Hayes JP (1978) Path complexity of logic networks. IEEE Trans Comput C-27: 459–462.

184. Nakagawa T (2002) Theoretical attempt of service reliability. J Reliab Eng Assoc Jpn 24: 259–260.

185. Nakamura S, Nakagawa T, Sandoh H (1998) Optimal number of spare cash-boxes for unmanned bank ATMs. RAIRO Oper Res 32: 389–398.

186. Nakamura S, Qian CH, Hayashi I, Nakagawa T (2003) An optimal maintenance time of automatic monitoring system of ATM with two kinds of breakdowns. Comput Math Appl 46: 1095–1101.

187. Nakamura S, Qian CH, Hayashi I, Nakagawa T (2002) Determination of loan interest rate considering bankruptcy and mortgage collection costs. Int Trans Oper Res 9: 695–701.

188. Arafuka M, Nakamura S, Nakagawa T, Kondo H (2007) Optimal interval of CRL issue in PKI architecture. In: Pham H (ed) Reliability Modeling Analysis and Optimization. World Scientific, Singapore: 67–79.

189. Nakamura S, Arafuka M, Nakagawa T (2007) Optimal certificate update interval considering communication costs in PKI. In: Dohi T, Osaki: S, Sawaki K (eds) Stochastic Operations Research. World Scientific, Singapore: 235–244.

190. Pham H (2000) Software Reliability. Springer, Singapore.

191. Pham H (2006) System Software Reliability. Springer, London.

192. Calabria R, Ragione LD, Pulcini G, Rapone M (1993) Service dependability of transit systems: A case study. In: Proceedings Annual Reliability and Maintainability Symposium: 366–371.

193. Masuda A (2003) A proposal of service reliability study and its practical application in maintenance support of electronics products. In: Proceedings Int IEEE Conference on the Business of Electronic Product Reliability and Liability: 119–125.

194. Wang N, Lu JC (2006) Reliability modeling in spatially distributed logistics systems. IEEE Trans Reliab 55: 525–534.

195. Johnson Jr AM, Malek M (1988) Survey of software tools for evaluating reliability, availability, and serviceability. ACM Comput Surveys 20: 227–269.

196. Chen W, Toueg S, Aguilera MK (2002) On the quality of service of failure detectors. IEEE Trans Comput 51: 561–580.

197. Dai YS, Xie M, Poh KL, Liu GQ (2003) A study of service reliability and availability for distributed systems. Reliab Eng Syst Saf 79: 103–112.

198. Levitin G, Dai YS (2007) Service reliability and performance in grid system with star topology. Reliab Eng Syst Saf 92: 40–46.

199. Scarf H, Gilford D, Shelly M (eds) (1963) Multistage Inventory Models and Techniques. Stanford University Press.

200. Sealey CW Jr (1980) Deposit rate-setting, risk aversion, and the theory of depository financial intermediaries. J Finance 35: 1139–1154.

201. Ho TSY, Saunders A (1981) The determinants of bank interest margins: Theory and empirical evidence. J Financial Quant Anal 16: 581-600.

202. Solvin MB, Sushka ME (1983) A model of commerial loan rate. J Finance 38: 1583–1596.

203. Allen L (1988) The determination of bank interest margins: A note. J Financial Quant Anal 23: 231–235.

204. Zarruk ER, Madura J (1992) Optimal bank interest margin under capital regulation and deposit insurance. J Financial Quant Anal 27: 143–149.
205. Angbazo L (1997) Commercial bank net interest margins, default risk, interest risk, and off-balance sheet banking. J Banking & Finance 21: 55–87.
206. Wong KP (1997) On the determinants of bank interest margins under credit and interest rate risks. J Banking & Finance 21: 251–271.
207. Athavale M, Edmister RO (1999) Borrowing relationships, monitoring, and the influence on loan rates. J Financial Res 22: 341–352.
208. Ebrahim MS, Mathur I (2000) Optimal entrepreneurial financial contracting. J Business Financial & Accounting 27: 1349–1374.
209. Michner M, Peterson RP (1957) An operations-research study of the collection of defaulted loans. Oper Res 5: 522–546.
210. Housley R, Ford W, Polk W, Solo D (1999) Internet X.509 public key infrastructure certificate and CRL profile. The Internet Society.
211. Cooper DA (2000) A more efficient use of Delta-CRLs. Proceeding of 2000 IEEE Symposium Security and Privacy: 190–202.
212. Chadwick DW, Young AJ (1997) Merging and extending the PGP and PEM trust models–The ICE-TEL trust model. IEEE Networks Special Publication on Network and Internet Security 11: 16–24.

Index